Synthese Library

Studies in Epistemology, Logic, Methodology,
and Philosophy of Science

Volume 427

Editor-in-Chief

Otávio Bueno, Department of Philosophy, University of Miami, Coral Gables, USA

Editorial Board Member

Berit Brogaard, University of Miami, Coral Gables, USA
Anjan Chakravartty, University of Notre Dame, Notre Dame, USA
Steven French, University of Leeds, Leeds, UK
Catarina Dutilh Novaes, VU Amsterdam, Amsterdam, The Netherlands
Darrell P. Rowbottom, Department of Philosophy, Lingnan University, Tuen Mun, Hong Kong
Emma Ruttkamp, Department of Philosophy, University of South Africa, Pretoria, South Africa
Kristie Miller, Department of Philosophy, Centre for Time, University of Sydney, Sydney, Australia

The aim of *Synthese Library* is to provide a forum for the best current work in the methodology and philosophy of science and in epistemology. A wide variety of different approaches have traditionally been represented in the Library, and every effort is made to maintain this variety, not for its own sake, but because we believe that there are many fruitful and illuminating approaches to the philosophy of science and related disciplines.

Special attention is paid to methodological studies which illustrate the interplay of empirical and philosophical viewpoints and to contributions to the formal (logical, set-theoretical, mathematical, information-theoretical, decision-theoretical, etc.) methodology of empirical sciences. Likewise, the applications of logical methods to epistemology as well as philosophically and methodologically relevant studies in logic are strongly encouraged. The emphasis on logic will be tempered by interest in the psychological, historical, and sociological aspects of science.

Besides monographs *Synthese Library* publishes thematically unified anthologies and edited volumes with a well-defined topical focus inside the aim and scope of the book series. The contributions in the volumes are expected to be focused and structurally organized in accordance with the central theme(s), and should be tied together by an extensive editorial introduction or set of introductions if the volume is divided into parts. An extensive bibliography and index are mandatory.

More information about this series at http://www.springer.com/series/6607

Roman Frigg • James Nguyen

Modelling Nature: An Opinionated Introduction to Scientific Representation

 Springer

Roman Frigg
Department of Philosophy, Logic
and Scientific Method
London School of Economics
& Political Science
London, UK

Centre for Philosophy of Natural
and Social Science
London School of Economics
& Political Science
London, UK

James Nguyen
Department of Philosophy
University College London
London, UK

Institute of Philosophy
University of London
London, UK

Centre for Philosophy of Natural
and Social Science
London School of Economics
& Political Science
London, UK

Synthese Library
ISBN 978-3-030-45152-3   ISBN 978-3-030-45153-0 (eBook)
https://doi.org/10.1007/978-3-030-45153-0

© Springer Nature Switzerland AG 2020

This work is subject to copyright. All rights are reserved by the Publisher, whether the whole or part of the material is concerned, specifically the rights of translation, reprinting, reuse of illustrations, recitation, broadcasting, reproduction on microfilms or in any other physical way, and transmission or information storage and retrieval, electronic adaptation, computer software, or by similar or dissimilar methodology now known or hereafter developed.

The use of general descriptive names, registered names, trademarks, service marks, etc. in this publication does not imply, even in the absence of a specific statement, that such names are exempt from the relevant protective laws and regulations and therefore free for general use.

The publisher, the authors, and the editors are safe to assume that the advice and information in this book are believed to be true and accurate at the date of publication. Neither the publisher nor the authors or the editors give a warranty, expressed or implied, with respect to the material contained herein or for any errors or omissions that may have been made. The publisher remains neutral with regard to jurisdictional claims in published maps and institutional affiliations.

This Springer imprint is published by the registered company Springer Nature Switzerland AG
The registered company address is: Gewerbestrasse 11, 6330 Cham, Switzerland

Preface

Models matter. Scientists spend much effort on constructing, improving, and testing models, and countless pages in scientific journals are filled with descriptions of models and their behaviours. Models owe much of their importance in the scientific process to the fact that many of them are representations, which allows scientists to study a model to discover features of reality. And the importance of representation is not limited to science. We look at photographs, contemplate paintings, study diagrams, read novels, watch movies, appreciate statues, are perplexed by kinematic installations, and watch the lights when crossing the road. There is hardly an aspect of our lives that is not permeated by representations. But what does it mean for something to represent something else? This is the question we discuss in this book. We focus on scientific representation, but, as we shall see, the boundaries between scientific representation and other kinds of representation are porous, if not spurious, and attempts to seperate scientific representation and analyse it in blissful isolation are doomed to failure.

The problem of scientific representation has by now generated a sizable literature, which has been growing particularly fast over the last decade. However, even a cursory look at this literature will leave the reader with the impression that the discussion about scientific representation is still in its infancy: there is no stable terminology, no shared understanding of what the central problems are, and no agreement on what might count as an acceptable solution. The aim of this book is threefold. Our first task is to get clear on what the problems are that we ought to come to grips with, how these problems should be formulated, and what criteria an acceptable solution has to satisfy. We then review the extant literature on the topic and assess the strengths and weaknesses of different proposals in the light of our conceptualisation of the problems and our criteria for adequate solutions. Finally, we offer our own answers to the quandaries of scientific representation and formulate what we call the DEKI account of representation.

Parts of the book build on previous publications. Chaps. 1, 2, 3, 4, and 5, and Sects. 7.1 and 7.2 are improved and expanded versions of our (2017a). We included new material in many places and updated the arguments in the light of criticisms and comments we received. Sect. 4.5 includes parts of our (2017); Sects. 7.3, 7.4,

7.5, and 7.6 are based on material from our (2017b); Chap. 8 includes material from our (2018); and Sects. 9.4 and 9.5 include material from our (2019a).

The book is intended to be intelligible to advanced undergraduate students, and it should also be useful for graduate seminars. We hope, however, that it will be of equal interest to professional philosophers and researchers in science studies, as well as to scientists and policy-makers who care about how, and what, models tell them about the world.

London, UK   Roman Frigg
                                                                                                       James Nguyen

Acknowledgements

In the years in which we have been thinking about representation, we have amassed a multitude of debts. Friends, colleagues, students, and the occasional inspired stranger helped to keep our arguments on track, move our thinking forward, and bring our musings to conclusions. Andreas Achen, Margherita Harris, David Lavis, Joe Roussos, and James Wills took it upon themselves to read the entire manuscript; Michal Hladky, Julian Reiss, Philippe Verreault-Julien, and Philipp Wichardt read chapters. Their considered comments and constructive suggestions have helped us improve the manuscript. For their guidance, insight, advice, and counsel since we started working on the manuscript, we would like to thank Joseph Berkovitz, Otávio Bueno, Jeremy Butterfield, Nancy Cartwright, José Díez, Catherine Elgin, Stacie Friend, Mathias Frisch, Manuel García-Carpintero, Peter Godfrey-Smith, Stephan Hartmann, Carl Hoefer, Laurenz Hudetz, Phyllis Illari, Elaine Landry, Sabina Leonelli, Arnon Levy, Genoveva Martí, Michela Massimi, Mary Morgan, Margaret Morrison, Fred Muller, Christopher Pincock, Demetris Portides, Stathis Psillos, Bryan W. Roberts, Leonard A. Smith, Fiora Salis, Nicholas J. Teh, David Teira, Paul Teller, Martin Thomson-Jones, Adam Toon, Michael Weisberg, and Martin Zach. Heartfelt apologies to anyone we have forgotten. Last, but not least, we would like to thank Otávio Bueno for suggesting to us that we embark on this book project, and for his continued encouragement throughout the process of writing. Without his support the project would never have seen the light of day.

We wish to acknowledge with gratitude support from the Alexander von Humboldt Foundation through a Friedrich Wilhelm Bessel Research Award (RF); the Arts and Humanities Research Council through a doctoral award (JN); the University of Notre Dame, the University of London, and University College London for postdoctoral support (JN); the British Academy for a Rising Star Engagement Award (JN); and the Jeffrey Rubinoff Sculpture Park (JN). Finally, RF wishes to thank Benedetta Rossi for being a wonderful companion and an insightful critic, and for her support during the process of writing the book – and for making it all worthwhile.

Introduction

Imagine you want to determine the orbit of a planet moving around the sun. You know that gravity pulls the planet and the sun towards each other and that their motion is governed by Newton's equation. To put this knowledge to use, you first have to construct a model of the system. You make the idealising assumption that the gravitational interaction between the sun and the planet is the only force relevant to the planet's motion, and you neglect all other forces, most notably the gravitational interaction between the planet and other objects in the universe. You furthermore assume that both the sun and the planet are perfect spheres with a homogenous mass distribution (meaning that the mass is evenly distributed within each sphere). This allows you to pretend that the gravitational interaction between the planet and the sun behaves as if the entire mass of each object were concentrated in its centre. Since the sun's mass is vastly larger than the mass of the planet, you assume that the sun is at rest and the planet orbits around it. With this model in place, you now turn to mechanics. Newton's equation of motion is $\vec{F} = m\vec{a}$, where $\vec{a}$ is the acceleration of a particle, m its mass, and $\vec{F}$ the force acting on it, and the law of gravity says that the magnitude of the force acting between the planet and the sun is $F_g = G\, m_p\, m_s/r^2$, where m_p and m_s are the masses of the planet and the sun, respectively, r is the distance between the two, and G is the gravitational constant. Placing the sun at the origin of the coordinate system and plugging F_g into the equation, you obtain $\ddot{\vec{x}} = -G m_s\, \vec{x}/|\vec{x}|^3$, where the double dots indicate the second derivative with respect to time. This is the differential equation describing the planet's trajectory, where you have, of course, used $\vec{a} = \ddot{\vec{x}}$, i.e. you utilised that acceleration is equal to the second derivative of position.

Constructing a model of the system has been crucial to deriving the desired result. In fact, without a model of the planet and the sun, you would not have been able to determine the planet's orbit. This example is not an exception. Models play a central role in science. Scientists construct models of atoms, elementary particles, polymers, populations, genetic trees, economies, rational decisions, aeroplanes, earthquakes, forest fires, irrigation systems, and the world's climate – there is hardly a domain of inquiry without models. Models are essential for the acquisition and organisation of scientific knowledge. So how do models work? How can it be the case that by studying a model, we can come to discover features of the thing that the model stands for? In this book we explore the idea that models work this way because they *represent* the selected parts or aspects of the world that we investigate. If we want to understand how models allow us to learn about the world, we have to come to understand how they represent.

Why is this important? Given the centrality of models in the scientific endeavour, the question of how models provide us with insight into the way the world is should concern anybody who is interested in understanding how science works. And given how central science is for understanding how we are situated in the world as epistemic agents – as agents who know things, who understand things, who categorise things, and so on – it should concern anybody who is interested in human cognitive endeavours. Furthermore, the question of how models represent is also conceptually prior to other debates concerning metaphysical, epistemological, and methodological questions in connection with science, and appropriate framings of these questions presuppose an understanding of how models represent.

The realism debate is a case in point. What does it mean to be a *scientific realist* about a model-based science? The usual way of characterising scientific realism is that mature scientific theories must be taken literally and be regarded as (approximately) true, both in what they say about observables and in what they say about unobservables (Psillos 1999). Despite many of the participants in this discussion rejecting a linguistic understanding of theories (associated with the so-called syntactic view of theories), the scientific realism debate is framed mostly in linguistic terms, focussing on the reference of theoretical terms and the (approximate) truth of theoretical statements. There is, at least on the face of it, a mismatch between an understanding of scientific theorising as an essentially model-based activity, which, as we will see, is not obviously linguistic in a straightforward sense, and the framing of the realism debate in linguistic terms (Chakravartty 2001). A reflection on how models represent can help us resolve this tension because it can help us understand what it means for models, or parts of models, to refer and to make truth-evaluable claims.[1]

The realism problem is often seen as particularly pressing in the context of model-based science because many models involve idealisations and approximations, or they are analogies of their targets. This has got enshrined in the categorisation of models, where it is common to classify models as idealised models, approximate models, or analogue models. This is salient in the current context because these classifications do not pertain to intrinsic features of models but to the ways in which models relate to their target systems. As such, idealisation, approximation, and analogy can be seen as being specific modes of representation, and a

---

[1] For recent discussions of scientific realism with a focus on models, see Reiss' (2012b) and Saatsi's (2016). For a general overview of models in science, see Bailer-Jones' (2002a) and Frigg and Hartmann's (2020). For a historical discussion of models in philosophy of science, see Bailer-Jones' (1999), and for a discussion of how physicists view their models, see her (2002b).

discussion of these modes might benefit from being situated in the wider context of a general theory of representation.[2]

Relatedly, how are we to understand scientific *pluralism*, or *perspectivism*, the idea that scientific practice provides us with multiple models of the same target system, either diachronically or synchronically? Are we to understand these multiple models as conflicting or complementary?[3] Again, this turns on how we understand their representational content.

Or consider the question of what it means for a model to *explain*. One popular way of analysing model-based explanation is to appeal to the idea that a model accurately captures the counterfactual profile of the target system because it either accurately represents how the target system would behave under various different conditions, or captures the difference makers of the phenomenon in question.[4] But this approach relies on us understanding how models can represent counterfactual behaviour, which requires an account of scientific representation. Further consider the notion that science provides us with *understanding* of features of the world.[5] This understanding is, at least in part, delivered by scientific models. But in order to know what it means for a model to provide understanding of a feature of the world, we have to have some grasp of the relationship between the model and the feature. And again, this relationship should be understood as a representational one.

So the question of scientific representation is foundational for various questions in the philosophy of science. This book is intended to provide those working on these questions, as well as those who are simply interested in the relationship between models and the world, with an introduction to the problem of scientific representation. Moreover, we hope that our discussion will be useful to scientists who are concerned with the relationship between their models and the aspects of the world that they are ultimately interested in. Beyond that, we hope that the book will be relevant for researchers in science studies interested in conceptual issues

---

[2] Recent discussions of idealisation and approximation with an angle on models can be found in Batterman's (2009), Jebeile and Kennedy's (2015), Nguyen's (2020), Norton's (2012), Portides' (2007), Potochnik's (2017), Saatsi's (2011a), and Vickers' (2016). For a recent discussion of analogue models, see Dardashti et al. (2017, 2019).

[3] There is a fast-growing literature on pluralism and perspectivism. For useful discussions, see Chakravartty's (2010), Chang's (2012), Giere's (2006), Massimi's (2017, 2018), Mitchell's (2002), Morrison's (2011), Rueger's (2005), Taylor and Vickers' (2017), and Teller's (2018), as well as the contributions to Massimi and McCoy's (2019).

[4] See, for instance, Bokulich's (2011) and Strevens' (2008). Again, the relationship between models and explanation is a significant issue in its own right. For more on the relationship between representation and explanation, see Lawler and Sullivan's (2020), Reiss' (2012a), and Woody's (2004).

[5] The question of scientific understanding, and the role models play in scientists' quest for understanding, has received increasing discussion in recent years. See, for instance, De Regt's (2017), Doyle et al. (2019), Elgin's (2004, 2017), Illari's (2019), Khalifa's (2017), Kostić's (2019), Le Bihan's (2019), Reutlinger et al. (2018), Sullivan and Khalifa's (2019), and Verreault-Julien's (2019), as well as the papers collected in Grimm et al. (2017).

concerning model-based science, philosophers working on topics related to representation, and policy-makers taking decisions based on model outputs.

Before delving into the details, two caveats are in order. Approaching scientific modelling by investigating representation is not an imperialist endeavour: our discussion is neither premised on the claim that *all* models are representational, nor does it assume that representation is the *only* function of models. It has been emphasised variously that models perform a number of functions other than representation. Knuuttila (2005, 2011) submits that the epistemic value of models is not limited to their representational function and develops an account that views models as epistemic artefacts that allow us to gather knowledge in diverse ways; Morgan and Morrison (1999) emphasise the role models play in the mediation between theories and the world; Hartmann (1995) and Leplin (1980) discuss models as tools for theory construction; Luczak (2017) talks about the non-representational roles played by toy models; Peschard (2011) investigates the way in which models may be used to construct other models and generate new target systems; Bokulich (2009) and Kennedy (2012) formulate non-representational accounts of model explanation;[6] and Isaac (2013) discusses nonexplanatory uses of models which do not rely on their representational capacities. Not only do we not see projects like these as being in conflict with a view that sees some models as representational; we think that the approaches are in fact complementary. Our point of departure is that *some* models represent and that therefore representation is *one of* the functions that these models perform. We believe that this is an important function and that it is therefore a worthy endeavour to enquire into how models manage to represent something beyond themselves.

The second caveat is that we are not presupposing that models are the *sole* unit of scientific representation, or that all scientific representation is model-based. Various types of images have their place in science, and so do graphs, diagrams, and drawings.[7] In some contexts, scientists also use what Warmbrōd (1992) calls "natural forms of representation" and what Peirce would have classified as indices, namely, signs that have a "direct physical connection" to what they signify (Hartshorne and Weiss 1931–1935, CP 1.372, cf. CP 2.92): tree rings, fingerprints, and disease symptoms. These are related to thermometer readings and litmus paper indications, which are commonly classified as measurements. Measurements also provide representations of processes in nature, sometimes together with the subsequent condensation of measurement results in the form of charts, curves, tables, and the like.[8] And, last, but not least, many would hold that theories represent too. At

---

[6] The issue of non-representational model explanations has also received attention phrased in terms of what Batterman and Rice (2014) call "minimal models". It is worth nothing, however, that the term is used in various ways in the literature. See, for instance, Fumagalli's (2015, 2016), Grüne-Yanoff's (2009, 2013), Jhun et al. (2018), and Weisberg's (2007).

[7] Downes (2012), Elkins (1999), and Perini (2005a, b, 2010) provide discussions of visual representation in science.

[8] Díez (1997a, b) and Tal (2017) offer general discussions of measurement. For a discussion of measurement in physics, in particular temperature, see Chang's (2004), and for a discussion of measurement in economics, see Reiss' (2001).

this point, the vexing problem of the nature of theories and the relation between theories and models rears its head again, and we refer the reader to Portides' (2017) for a discussion of this issue.

There is no question that these forms of "non-model representation" exist – they do and they play important roles in various branches of science. The question is whether these other kinds of representation function in ways that are fundamentally different from the way in which models function. Do, say, graphs represent in the same way that models do? The answer to this question will depend on what one has to say about models and hence depends on one's account of representation. What all accounts of scientific representation have in common is that they must address the issue. An account of scientific representation remains incomplete as long as it does not specify how it deals with alternative forms of representation.

The book is organised as follows. In Chap. 1 we reflect on the tasks ahead and present a list with five problems that every account of representation must answer, along with five conditions of adequacy that every viable answer must meet. These questions and conditions provide the analytical lens through which we look at the different accounts of representation in subsequent chapters.[9] In Chap. 2 we discuss Griceanism and representation by stipulation: the claim that models represent their targets because we intend them to, and that's all there is to say about the matter. In Chap. 3 we look at the time-honoured similarity approach, and in Chap. 4 we examine its modern-day cousin, the structuralist approach. Both, in relevantly different ways, take similarities, structural or otherwise, between models and their targets to be constitutive of scientific representation. In Chap. 5 we turn to inferentialism, a more recent family of conceptions which emphasise the role that models play in generating hypotheses about their targets. In Chap. 6 we discuss the fiction view of models and distinguish between different versions of the view. In Chap. 7 we consider accounts based on the notion of "representation-as", which identify the fact that models represent their subject matter as being thus or so as the core of a theory of representation.

While this book is an introduction to the literature, and while we have endeavoured to provide a balanced treatment of the positions we discuss, the book is also, as indicated in its title, an *opinionated* introduction. The conclusion we reach at the end of Chap. 7 is that all currently available positions are beset with problems and that a novel approach is required. This is our project in the final two chapters of the book. In Chap. 8 we develop what we call the DEKI account of representation and explain how it works in the context of material models. In Chap. 9 we generalise the DEKI account to ensure it applies to non-material models, and reflect on the relation between representation in art and science.

---

[9] A historical introduction to the issue of scientific representation can be found in Boniolo's (2007).

Contents

1 Problems Concerning Scientific Representation 1
   1.1 Aboutness, Surrogative Reasoning, and Demarcation........... 2
   1.2 The ER-Scheme and Representational Styles 9
   1.3 Accuracy, Misrepresentation, and Targetless Models 12
   1.4 The Application of Mathematics 15
   1.5 The Carriers of Representations......................... 15
   1.6 Chapter Summary 18

2 General Griceanism and Stipulative Fiat...................... 23
   2.1 Demarcation and the ER-Problem........................ 23
   2.2 Taking Stock 26
   2.3 General Griceanism and Stipulative Fiat.................... 27

3 The Similarity View 31
   3.1 The Demarcation Problems 32
   3.2 The Indirect Epistemic Representation Problem 35
   3.3 Accuracy and Style 42
   3.4 The Problem of Carriers................................ 48

4 The Structuralist View 51
   4.1 The Demarcation Problems 53
   4.2 The Problem of Carriers................................ 56
   4.3 The Indirect Epistemic Representation Problem 60
   4.4 Accuracy and Style 64
   4.5 The Structure of Target Systems 70

5 The Inferential View 83
   5.1 The Demarcation Problems 83
   5.2 Deflationary Inferentialism 85
   5.3 Motivating Deflationism 87
   5.4 Reinflating Inferentialism: Interpretation 95

6	**The Fiction View of Models**	105
	6.1 Intuitions and Questions	106
	6.2 Two Notions of Fiction	109
	6.3 Fiction and the Problem of Carriers	114
	6.4 A Primer on Pretence Theory	119
	6.5 Models and Make-Believe	121
	6.6 Make-Believe, Ontology and Directionality	123
	6.7 Direct Representation	126
	6.8 Against Fiction	131
7	**Representation-As**	137
	7.1 Demarcation	138
	7.2 Denotation, Demonstration, and Interpretation	139
	7.3 Denotation and Z-Representation	143
	7.4 Exemplification	147
	7.5 Defining Representation-As	149
	7.6 Unfinished Business	152
8	**The DEKI Account**	159
	8.1 Water Pumps, Intuition Pumps	160
	8.2 Models and Z-Representations	166
	8.3 Exemplification Revisited	172
	8.4 Imputation and Keys	174
	8.5 The DEKI Account of Representation	176
	8.6 Reverberations	181
9	**DEKI Goes Forth**	185
	9.1 DEKI with Non-material Models	185
	9.2 The Applicability of Mathematics	191
	9.3 Limit Keys	195
	9.4 Beyond Models	204
	9.5 Difference in Sameness	210

Bibliography ... 215

Name Index ... 233

Subject Index ... 239

Chapter 1
Problems Concerning Scientific Representation

What questions does a philosophical account of scientific representation have to answer and what conditions do these answers have to satisfy? Different authors have focussed on different issues and framed the problem in different ways. As we noted in the Preface, there is neither a shared understanding of the problems that an account of scientific representation has to address; nor is there agreement on what an acceptable solution to these problems would look like. In fact, there is not even a stable and standard terminology in which problems can be formulated. The aim of this chapter is to develop such a terminology, to state what we take to be the core problems that every account of representation must solve, and to formulate conditions of adequacy that acceptable solutions to these problems have to meet.[1] This leads us to identify five problems and five conditions. These will be used to structure the discussion throughout the book and to evaluate views and positions. It is a coincidence that the number of conditions is equal to the number of questions. The conditions of adequacy are not "paired up" with the questions (so that, for instance, the first condition would concern the first question, and so on). The conditions are independent of particular questions.

We begin our discussion by looking at a representation's "aboutness", its ability to support surrogative reasoning, and the problem of demarcating different sorts of representation (Sect. 1.1). Not all representations are of the same kind and the same target can be represented in different styles (Sect. 1.2). Some representations are accurate while others are misrepresentations, and some models have no target system at all despite being representations (Sect. 1.3). Many representations are mathematised, which raises the question of how mathematics hooks onto something in the physical world (Sect. 1.4). The last question concerns the nature of the objects

---

[1] To frame the issue of representation in terms of five problems is not to say that these are separate and unrelated issues that can be dealt with one after the other in roughly the same way in which we first buy a ticket, then walk to the platform, and finally take a train. The division is analytical, not factual.

that do the representing: what kind of objects are models and how are they handled in scientific practice (Sect. 1.5)?

The discussion in these sections is dense, and it is easy to lose track of the various points. For this reason we end the chapter with a summary (Sect. 1.6). The summary serves two purposes. The first, and obvious, purpose is to recapitulate, in concise form, the different problems and conditions that will guide our discussion, and thereby provide an easily accessible point of reference for later debates. The second, and less obvious, purpose is to offer the hurried reader a convenient way forward. We realise that the discussions in Sect. 1.1 through to Sect. 1.5 go into more detail than some readers may care to engage with. The summary in Sect. 1.6 is self-contained and understandable even when read in isolation. Those who prefer to bypass lengthy philosophical reflections concerning the nature of a theory of representation and wish to get into a discussion of the different accounts of representation straightaway can now fast-forward to Sect. 1.6 *without* first reading the other sections in this chapter.

1.1 Aboutness, Surrogative Reasoning, and Demarcation

The selected part or aspect of the world that is represented by a representation is the representation's *target system* (or *target*, for short). The target of a portrait of Newton is Sir Issac Newton; the target of his model is the solar system. The object that is doing the representing is the *carrier*.[2] A canvas covered with pigments is the carrier of Newton's portrait, and the object described at the beginning of the Introduction is the carrier of the model of the solar system. Representation, then, is the relation between a carrier and a target. We follow common usage and speak of "a representation" when we refer to the carrier *as related to* its target. In this sense Newton's portrait is a representation, and the model of the solar system is a representation. An *account of representation* offers answers to questions that arise in connection with representation, and *conditions of adequacy* state requirements that a satisfactory account of representation must satisfy.

Models have "aboutness": they are representations *of* a target system. The first and most fundamental question concerning scientific representation therefore is: in virtue of what is a model a scientific representation of something else?[3] To appreciate the thrust of this question it is instructive to briefly consider the parallel problem in the context of pictorial representation. When seeing, say, van Gough's *Self-Portrait with Bandaged Ear* we immediately realise that it depicts a man wearing a green jacket and a blue hat. Why is this? Symbolist painter Maurice Denis issued

---

[2] Contessa speaks of the "vehicle" of a representation (2007, p. 51) to cover all objects that are used by someone to represent something else, and Suárez prefers the term "source" (2004, p. 768). While both are workable suggestions, "vehicle" and "source" are now too closely associated with their respective accounts of representation to serve as a neutral term to state our problem.

[3] Explicit attention has been drawn to this question by Frigg (2002, p. 2, 17; 2006, p. 50), Morrison (2008, p. 70), Stachowiak (1973, p. 131), and Suárez (2003, p. 230).

1.1 Aboutness, Surrogative Reasoning, and Demarcation

the by now notorious warning to his fellow artists that "[w]e should remember that a picture – before being a war horse, a nude woman, or telling some other story – is essentially a flat surface covered with colours arranged in a particular pattern" (Denis 2008, p. 863). This gets right to the point: how does an arrangement of pigments on a surface represent something outside the picture frame?

There is nothing peculiar about pictures and the same question arises for every representation: what turns something into a representation of something else? We call this the *Representation Problem*. This problem comes in different versions, and, as we will see, which version one choses to address will depend on one's views concerning different kinds of representations. An obvious kind of representation to investigate, at least in the current context, is scientific representation (we come to other kinds soon). Echoing Denis' remark about pictures we can then say that before being a representation of an atom, a population, or an economy, a model is an equation, a structure, a fictional scenario, or a mannerly physical object. What turns mathematical equations or structures, or fictional scenarios or physical objects, into representations of something beyond themselves? It has become customary to phrase this problem in terms of necessary and sufficient conditions. The question then is: what fills the blank in "X is a scientific representation of T iff ___", where "X" stands for the carrier, and "T" the target system? We call this the *Scientific Representation Problem* (*SR-Problem*, for short), and the biconditional the *SR-Scheme*. So one can say that the SR-Problem is to fill the blank in the SR-Scheme.

A central feature of representation is its directionality. A portrait represents its subject, but the subject does not represent the portrait. Likewise, a model is about its target, but the target is not about the model. An account of representation has to identify the root of this directionality, and this means that the SR-Problem has to be solved in way that accounts for it. We call this the *Condition of Directionality*, which is our first condition of adequacy for an account of representation.

To spare ourselves difficulties further down the line, our formulation of the Representation Problem needs to be qualified in the light of a second condition of adequacy that any account of scientific representation has to meet. The condition is that models represent in a way that allows scientists to form hypotheses about the models' target systems: they can generate claims about target systems by investigating models that represent them. As we have seen in the Introduction, students of mechanics can generate claims about the trajectory of a planet by studying the properties of the Newtonian model. This is no exception. Many investigations are carried out on models rather than on reality itself, and this is done with the aim of discovering features of the things that models stand for. Every acceptable theory of scientific representation has to account for how reasoning conducted on models can yield claims about their target systems, and there seems to be widespread agreement on this point.[4] Following Swoyer (1991, p. 449) we call this the *Surrogative Reasoning Condition*.

---

[4] Bailer-Jones (2003, p. 59, origional emphasis) notes that models "*tell us* something about certain features of the world"; Bolinska (2013) and Contessa (2007) both call models "epistemic representations"; Frigg (2003, p. 104; 2006, p. 51) sees the potential for learning as an essential explanan-

This gives rise to a potential problem for the SR-Scheme. The problem is that any account of representation that fills the blank in that scheme in a way that satisfies the Surrogative Reasoning Condition will almost invariably end up covering kinds of representation that one may not want to qualify as scientific representations. Pictures, photographs, maps, diagrams, charts, and drawings, among others, often provide epistemic access to features of the things they represent, and hence they may fall under an account of representation that explains surrogative reasoning. However, at least some of them are not scientific representations. While the photograph of a cell taken with a microscope and a chart of the temperature in the test reactor can be regarded as scientific representations in a broader sense, the portrait of the leader of an expedition, a photograph of the Houses of Parliament, and a drawing of a ballet scene are not scientific representations and yet they provide epistemic access to their subject matter in one way or another. This is a problem for an analysis of scientific representation in terms of necessary and sufficient conditions because if a representation that is not prima facie a scientific representation (for instance a portrait) satisfies the conditions of an account of scientific representation, then one either has to conclude that the account fails because it does not provide sufficient conditions,[5] or that first impressions are wrong and that the representation actually is a scientific representation.

Neither of these options is appealing. To avoid this problem one can follow a suggestion of Contessa's (2007) and broaden the scope of the investigation. Rather than analysing the category of scientific representation, one can analyse the broader category of *epistemic representation*. This category comprises models, but it also includes all other kinds of representation that allow for surrogative reasoning such as maps, photographs, and diagrams. The task then becomes to fill the blank in "X is an epistemic representation of T iff ___", where, again, "X" stands for the carrier and "T" for the target system. For brevity we use "$R(X, T)$" as a stand-in for "X is an epistemic representation of T", and so the biconditional becomes "$R(X, T)$ iff ___". We call the general problem of figuring out in virtue of what something is an epistemic representation of something else the *Epistemic Representation Problem* (*ER-Problem*, for short), and the biconditional "$R(X, T)$ iff ___" the *ER-Scheme*. The ER-Problem then is to fill the blank in the ER-Scheme.[6]

---

dum for any theory of representation; Liu (2013, p. 93) emphasises that the main role for models in science and technology is epistemic; Morgan and Morrison regard models as "investigative tools" (1999, p 11); Poznic says that "studying models is to pursue an epistemic purpose: modelers want to learn something" and thereby "gain insight into target systems that are represented by the models" (2016a, p. 202); Suárez (2003, p. 229; 2004, p. 772) submits that models licence specific inferences about their targets; and Weisberg (2013, p. 150) observes that the "model-world relation is the relationship in virtue of which studying a model can tell us something about the nature of a target system".

[5] We nuance this in Sect. 1.2, where we qualify the role of necessary and sufficient conditions.

[6] Frigg (2006, p. 50) calls this the "enigma of representation" and in Suárez's (2003, p. 230) terminology this amounts to identifying the "constituents" of a representation. In his (2004) Suárez explicitly offers only necessary but insufficient conditions on M representing T. Although it seems

1.1 Aboutness, Surrogative Reasoning, and Demarcation

The question of whether one should address the SR-Problem or the ER-Problem gives rise to the *Scientific Representational Demarcation Problem*: do scientific representations differ from other kinds of epistemic representations, and, if so, wherein does the difference lie?[7] It is important to note that the Scientific Representational Demarcation Problem concerns the question whether there is a difference between scientific representations and other kinds of epistemic representations *as regards their representational characteristics*. There may be any number of other differences. Scientific representations characteristically are produced by people who call themselves scientists, are published in scientific journals, are discussed at scientific conferences, and so on; non-scientific epistemic representations typically do not have these features. However, considerations pertaining to the history of production and social function of representations are irrelevant to the Scientific Representational Demarcation Problem (or at least are only relevant to the extent that they are relevant to their representational characteristics).

Those who give a positive answer to the Scientific Representational Demarcation Problem and therefore maintain that scientific representations have to be demarcated will, in the first instance, have to offer a solution to the SR-Problem. They may then address the ER-problem and show what sets scientific representations apart from other epistemic representations. Those who give a negative answer believe that scientific representations are not fundamentally different from other epistemic representations and can therefore turn to the ER-problem right away.

At this point a second demarcation problem arises. As noted in the Introduction, science employs representations that are not usually deemed models: theories, scientific images, graphs, diagrams, and so on. And the variety of types of representation would seem to be even larger in artistic contexts, where one finds paintings, drawings, etchings, sculptures, video installations, and many more. This gives rise to the *Taxonomic Representational Demarcation Problem*: are there different types of representations, and, if so, what are these types and wherein do the differences between them lie? As in the case of the Scientific Representational Demarcation Problem, the Taxonomic Representational Demarcation Problem only concerns the question whether there is a difference between different types of representations *as regards their representational characteristics*. Hence, even those who give a negative answer to the question are not forced to say that all representations are the same, or that all scientific representations are models. They are only committed to saying that there is no difference between different types of representations in the *way in which they represent*, and that any differences that one may identify are external to issues of representation.

As their names suggest, we understand the *Scientific Representational Demarcation Problem* and the *Taxonomic Representational Demarcation Problem* as two subproblems of the same overarching problem, the *Representational Demarcation Problem*,

---

like the ER-Scheme rules out his account from the offset, in Chap. 5 we argue that a plausible reading of the inferential conception fits neatly into the ER-scheme.

[7] Callender and Cohen discuss what they call "a kind of demarcation problem for scientific representation" (2006, pp. 68–69), which is effectively this problem.

the question whether there are different types of representations and, if so, what the types are.

The two demarcation problems are independent of each other in that it is possible to give a negative answer to one while giving a positive answer to the other. We call someone who gives negative answers to both a *universalist* because this denial amounts to believing that all epistemic representations function in the same way (representationally speaking), and that they are therefore covered by the same account. A universalist will address the ER-Problem.

Someone who denies taxonomic demarcation while upholding scientific demarcation believes that there is a difference between scientific and non-scientific representations but that otherwise representations, both scientific and non-scientific are the same (again, as far as their representational function is concerned). Someone who holds this view will address the SR-Problem.

Those who buy into taxonomical demarcation will have to say what types of representation they recognise, and each type Y will then require its own analysis with the aim of filling the blank in "X is a scientific/epistemic representation type Y of T iff ___", where "X" and "T" are as above. Either "scientific" or "epistemic" is chosen depending on whether they also buy into scientific demarcation: someone who does will choose "scientific", while someone who rejects scientific demarcation will choose "epistemic". We call this the *Type Representation Problem* (*TR-Problem*, for short), and the corresponding biconditional is the *TR-Scheme*. Addressing the TR-Problem then amounts to first making a list of all types of representation that one recognises and then completing the TR-Scheme for each type.

What would a taxonomic demarcation look like? Since models are of central importance in our context, one would expect that they would be among the recognised representational types, and that there would be a philosophical discussion around a "model representation problem". This problem would consist in filling the blank in the TR-Scheme for models, namely "X is a scientific/epistemic model representation of T iff ___". Interestingly, this is not the turn that the discussion has taken. In fact, models do not seem to get recognised as representational types sui generis in the sense that those who discuss "how models represent", by and large, discuss models in tandem with other kinds of epistemic representations. As we will see in Chap. 3, the taxonomic demarcation that has taken hold in the discussion is the distinction between direct and indirect representations. It so happens that proponents of this distinction reject scientific demarcation, and so they will aim to fill the blank in two TR-Schemes for epistemic representation, namely "X is a direct epistemic representation of T iff ___" and "X is an indirect epistemic representation of T iff ___".

The different versions of the Representation Problem and their dependence on the answers to the two demarcation problems are summarised in the matrix in Fig. 1.1. It is worth spelling out these options in a more detail. Someone who denies scientific demarcation while upholding taxonomic demarcation believes that there are different kinds of epistemic representations and that these require separate analyses, but at the same time believes that these kinds cut across the science versus

1.1 Aboutness, Surrogative Reasoning, and Demarcation

	Taxonomic Representational Demarcation Problem	
	Yes	No
Scientific Representational Demarcation Problem — Yes	Address the Type Representation Problem for "scientific representation"	Address the Scientific Representation Problem
Scientific Representational Demarcation Problem — No	Address the Type Representation Problem for "epistemic representation"	Address the Epistemic Representation Problem

Fig. 1.1 Matrix showing how different answers to the Scientific Representational Demarcation Problem and the Taxonomic Representational Demarcation Problem determine which version of the Representation Problem one is going to address

non-science divide in that scientific and non-scientific uses of these types are covered by the same philosophical analysis. On such a view there are, say, scientific as well as non-scientific photographs, and scientific as well as non-scientific models, but these are covered by the same analysis.[8] Proponents will therefore address the TR-Problem for "epistemic representation".

Someone who responds positively to both demarcation problems has to distinguish between different kinds of representation both within the class of scientific representations and the class of non-scientific representations. A theorist of this sort has two options. She might think that even though there is a difference between scientific and non-scientific representations, the internal division of both groups are the same. On such a view there could, for instance, be scientific and non-scientific photographs, scientific and non-scientific diagrams, scientific and non-scientific models, and so on, but the scientific version of a type would have to be covered by a different analysis than its non-scientific cousin. Another option is to think that subdivisions of scientific and non-scientific representations are different to begin with, which means that each group requires its own taxonomical scheme.

The original demarcation problem – to distinguish between science and non-science, and in particular pseudo-science – is not as active a research problem as it once was. For this reason, those interested in representation may also have little

---

[8] The notion of a non-scientific model is not an artefact of our classification. That the use of models need not be confined to science becomes clear from the recent debate about the use of models within philosophy. For a discussion of models in various parts of philosophy see, for instance, Colyvan's (2013), Godfrey-Smith's (2012), Hartmann's (2008), Sprenger and Hartmann's (2019, Chap. 1), and Williamson's (2018).

enthusiasm to get involved in questions of demarcation. Accordingly, with the exception of Callender and Cohen, who note their lack of optimism about solving this problem (2006, p. 83), the Scientific Representational Demarcation Problem seems to have received little, if any, explicit attention in the recent literature on scientific representation. This can be seen as suggesting that authors favour a negative answer. Indeed, such an answer seems to be implicit in approaches that discuss scientific representation alongside other forms of epistemic representation such as pictorial representation. Elgin's (2010), French's (2003), Frigg's (2006), Suárez's (2004), and van Fraassen's (2008) are examples of approaches of this kind.

However, the two representational demarcation problems are of systematic importance even if one ends up not demarcating at all. This is because, as have just seen, they have implications for how we address the problem of representation and what examples we use. Those who give negative answers to both problems can address the ER-problem directly and do not have to draw other distinctions. Those who give positive answers to at least one of the demarcation problems will have to address different representation problems and face different possible counterexamples. So one's stance on demarcation has clear methodological implications, and even those who are not inclined to engage with the problem at any level of detail will have to make their choices explicit to avoid causing difficulties downstream.

Our study of the issues concerning representation is not yet complete, and having put a tripartite distinction between the ER-Problem, the SR-Problem, and the different TR-Problems into place does, strictly speaking, require us to address all further issues that we encounter with respect to all three problems. We refrain from doing so and focus on the ER-Problem for three reasons. First, carrying all problems with us would lead to meandering strings of distinctions that would be hard to keep track of even for veterans in matters of representation. Second, it would fill pages with redundancies because the points we make about the ER-scheme carry over the other schemes mutatis mutandis. Third, while it is important for our analysis that we are clear on what problems there are and which of the problems we address, most positions in the current debate are either universalist, or nearly universalist in that they uphold only few demarcations. Even though discussions have focussed on models, this seems to be a pragmatic decision based on the fact that models are important. But for universalists (who reject the demarcation), at bottom all instances of epistemic representation function in the same way and the question whether the analysis starts with models, pictures, or yet something else, loses its teeth because any starting point will lead to the same result. A universalist studies models as *samples* of epistemic representation and claims that the result of the study generalises. We adopt this strategy in what follows and discuss models as instances of epistemic representation. Those who disagree with this classification can retrace the previous steps and re-introduce the distinctions we now suppress.

1.2 The ER-Scheme and Representational Styles

Even someone who accepts both representational demarcation problems and distinguishes between different types of representation will grant that not all representations of the same type work in the same way.[9] Consider the example of (non-scientific) painting, where the point is so obvious that it hardly needs mention: an Egyptian mural, a two-point perspective aquarelle, and a pointillist oil painting represent their respective targets in different ways. This pluralism is not limited to visual representations. As we have previously seen, there seem to be various types of scientific representation. And even if we restrict our focus to models, they don't all function in the same way. For example, Woody (2000) argues that chemistry as a discipline has its own ways to represent molecules. But differences in style can even appear in models from the same discipline. Weizsäcker's liquid drop model represents the nucleus of an atom in a manner that seems to be different from the one of the shell model. A scale model of the wing of a plane represents the wing in a way that is different from how a mathematical model of its cross section does. Or Phillips and Newlyn's famous hydraulic machine and Hicks' mathematical equations both represent a Keynesian economy but they seem to do so in different ways. In other words: they employ different *styles*. This gives rise to the question: what styles are there and how can they be characterised? This is the *Problem of Style* (Frigg 2006, p. 50). There is no expectation that a *complete* list of styles be provided in response to this problem. Indeed, it is unlikely that such a list can ever be drawn up, and new styles will be invented as science progresses. A response to the problem of style will always be open-ended, providing a taxonomy of what is currently available while leaving room for new additions.

How can different styles be accommodated in the ER-Scheme? One might worry that the scheme seems to assume that epistemic representation is a monolithic concept and thereby make it impossible to distinguish between different kinds of representation. This impression is engendered by the fact the scheme asks us to fill a blank, and blank is filled only once. But if there are different styles of representation, we should be able to fill the blank in different ways on different occasions. The answer to this problem lies in the realisation that the ER-Scheme is more flexible than appears at first sight. In fact, there are at least three different ways in which different styles of representations can be accommodated in the ER-Scheme. To pinpoint the locus of flexibility let us replace the blank with variable for a condition on representation and rewrite the ER-Scheme as "$R(X, T)$ iff C", where C denotes a condition (and, as we have seen in the previous section, $R(X, T)$ is a stand-in for "X is an epistemic representation of T"). The scheme then says that X is an epistemic representation of T iff C obtains. The condition will usually involve a relation between X and T (although, as we will see shortly, there can also be other relata). For this reason we call C the *grounding relation* of an epistemic representation. The ER-Problem now is to identify C.

---

[9] For someone who does not demarcate, or only demarcates along one dimension but not the other, the observation that there are different representational styles is even more obvious.

To see how the introduction of a grounding relation helps us to deal with styles, let us assume, for the sake of illustration, that we have identified two styles: analogue representation and idealised representation. The result of an analysis of these relations is the identification of their respective grounding relations, C_A and C_I. The first way of accommodating them in the ER-Scheme is to fill the blank with the disjunction of the two: "$R(X, T)$ iff C_A or C_I". In plain English: X is an epistemic representation of T if and only if X is an analogue representation of T or X is an idealised representation of T. This move is possible because, first appearances notwithstanding, nothing hangs on the grounding relation being homogeneous. The relation can be as complicated as we like and there is no prohibition against disjunctions. In the above case we have $C = [C_A \text{ or } C_I]$. The grounding relation could even be an open disjunction. This would help accommodate the observation that a list of styles is potentially open-ended. In that case there would be a grounding relation for each style and the scheme could be written as "$R(X, T)$ iff C_1 or C_2 or C_3 or ...", where the C_i are the grounding relations for different styles. This is not a new scheme; it's the same old scheme where $C = [C_1 \text{ or } C_2 \text{ or } C_3 \text{ or } ...]$ is spelled out.

Alternatively, one could formulate a different scheme for every kind of representation. This would amount to changing the scheme slightly in that one would then no longer analyse epistemic representation per se. Instead one would analyse different styles of epistemic representations. Consider the above example again. The response to the ER-Problem then consists in presenting the two biconditionals "$R_A(X, T)$ iff C_A" and "$R_I(X, T)$ iff C_I", where $R_A(X, T)$ stands for "X is an analogical epistemic representation of T" and $R_I(X, T)$ for "X is an idealised epistemic representation of T". This generalises straightforwardly to the case of any number of styles, and the open-endedness of the list of styles can be reflected in the fact that an open-ended list of conditionals of the form "$R_i(X, T)$ iff C_i" can be given, where the index ranges over styles.

In contrast with the second option, which pulls in the direction of more heterogeneity, the third option aims for more unity. The crucial observation here is that the grounding relation can in principle be an abstract relation that can be concretised in different ways, or a determinable that can have different determinates. On the third view, then, the concept of epistemic representation is like the concept of force (which is abstract in that in a concrete situation the acting force is gravity, or electromagnetic attraction, or some other specific force), or like colour (where a coloured object must be blue, or green, or ...). This view would leave "$R(X, T)$ iff C" unchanged and take it as understood that C is an abstract relation.

At this point we do not adjudicate between these options. Each has its pros and cons, and which is the most convenient to work with depends on one's other philosophical commitments. What matters is that the ER-scheme does have the flexibility to accommodate different representational styles, and that it can accommodate them in at least three different ways.[10]

---

[10] In passing we note that these accommodations can be done at all levels of analysis and hence can also be used to clarify the issues of demarcation discussed previously. The scheme's flexibility allows for the option of demarcating between scientific and non-scientific epistemic representa-

1.2 The ER-Scheme and Representational Styles

What kind of relation is the grounding relation? Some might worry that formulating the problem in this way presupposes that representation is an intrinsic relation between X and T: a relation that only depends on intrinsic properties of X and T and on how they relate to one another. That is, the worry is that the relation has the form $C(X, T)$, which precludes other factors from playing a role in the grounding relation. This is not so. In fact, R might depend on any number of factors other than X and T themselves. To make this explicit we can write the ER-Scheme in the form "$R(X, T)$ iff $C(X, T, x_1, \ldots, x_n)$", where n is a natural number and C is an $(n + 2)$-ary relation that grounds epistemic representation. The x_i can be anything that is deemed relevant to epistemic representation, for instance a user's intentions, standards of accuracy, and specific purposes.

This is the place to address another worry about the form of the ER-Scheme, namely that it is phrased in terms of necessary and sufficient conditions. Analysing a concept by attempting to offer such conditions has received some bad press in other areas of philosophy, mostly because it presupposes that concepts have sharp boundaries while in reality these boundaries are vague and porous. Therefore, so the argument goes, a project that attempts to offer necessary and sufficient condition for broad concepts like *scientific representation* or *epistemic representation* is starting on the wrong foot.[11]

If the primary aim of our project were to delineate the exact boundaries of these concepts, then the reasons for rejecting an analysis of concepts based on necessary and sufficient conditions would be pertinent. But our project is a different one. Rather than paying attention to the problem of exact delineation and "boundary cases" – the sorts of cases that usually plague attempts to define a concept – we focus on paradigmatic cases (or stereotypes, as they are often called in this context). Unlike in other cases where it is clear how to understand paradigmatic examples and controversies revolve around borderline cases, in the context of thinking about epistemic representation (or any other notion of representation we have introduced) there does not seem to be any widespread agreement about how the paradigmatic instances work. Each of the accounts we discuss in this book explicates paradigmatic examples in different ways, and we must address paradigmatic cases before getting worried about boundary cases. Goodman's methodological advice about how to analyse counterfactuals is pertinent also to our analysis of epistemic representation: "if we can provide an interpretation that handles the clear cases successfully, we can let the unclear ones fall where they may" (1983, p. xviii). Our point then is that, in the first instance at least, analysing epistemic representation in terms of necessary and sufficient conditions will allow us to get a handle on the role of the important features of paradigmatic cases, and once we understand those, an additional debate can be had as to the correct way of analysing borderline cases. Phrasing the problem in terms of necessary and sufficient conditions provides a clear

---

tion, and/or providing a taxonomy thereof, each of which can be analysed in any of the three manners discussed.

[11] For a general discussion of how to analyse concepts and the problems faced by approaches based on necessary and sufficient conditions see Laurence and Margolis' (1999).

framework in which to compare different ways of addressing the problem and furthers our understanding of how paradigmatic cases work.

1.3 Accuracy, Misrepresentation, and Targetless Models

The next problem for a theory of epistemic representation is to specify standards of accuracy. Some representations are accurate; others aren't. The Schrödinger model is an accurate representation of the hydrogen atom; the Thomson "plum-pudding" model isn't. On what grounds do we make such judgments? In Morrison's words: "how do we identify what constitutes an accurate representation?" (2008, p. 70). We call this the *Problem of Accuracy*. Accuracy allows for degrees and is sensitive to the purposes of a model and the context in which it is used.[12] Providing a response to this problem is a crucial aspect of an account of epistemic representation.

Being an accurate representation is not tantamount to being a mirror image, despite the popular myth that accurate epistemic representations should mirror, copy, or imitate the things they represent. On such a view, accurate representation is *ipso facto* realistic representation, but this is a mistake. Accurate representations can be realistic, but they need not.[13] And an accurate representation certainly need not be a copy of the real thing. A road map, for instance, can be an accurate representation of a country's highway system without thereby being a copy of it. Throughout this book, we encounter positions that make room for representations that aren't mirror images, and these positions testify to the fact that accurate representation is a much broader notion than mirroring.

Providing standards of accuracy goes hand in hand with a third condition of adequacy: the possibility of misrepresentation. Asking what makes a representation an accurate representation already presupposes that inaccurate representations are representations too. And this is how it should be. If X does not accurately portray T, then it is a misrepresentation but not a non-representation. It is therefore a general constraint on a theory of epistemic representation that it has to make misrepresentation possible.[14] We call this the *Misrepresentation Condition*. This condition can be motivated by a brief glance at the history of science, which is rife with inaccurate representations. Some of these were known to be inaccurate all along (like the

---

[12] See Parker's (2020) for a discussion about the relation between the purposes of using a model and standards of accuracy.

[13] This, we take it, is the moral of the satire about the cartographers who produce maps as large as the country itself just then see them abandoned. The story has been told by Lewis Carroll in *Sylvie and Bruno* and Jorge Luis Borges in *On Exactitude in Science*.

[14] This condition is adapted from Stich and Warfield (1994, pp. 6–7), who suggest that a theory of mental representation should be able to account for misrepresentation, as do Sterelny and Griffiths (1999, p. 104) in their discussion of genetic representation. The point has also been made by Frigg (2002, p. 16) and Suárez (2003, p. 235).

spherical planets in Newton's model of the solar system); others were, at the time, taken to be accurate and turned out to be inaccurate with hindsight (like Thomson's "plum pudding" model of the atom). A notion of representation that rules out the possibility of misrepresentation is defective. A corollary of this requirement is that representation is a wider concept than accurate representation and that representation cannot be analysed solely in terms of accurate representation.[15]

A related condition concerns models that misrepresent in the sense that they lack a target system altogether. We call such models *targetless models*. Models of four-sex organism populations are targetless because there are no species that have four different sexes that mate to produce offspring (Weisberg 2012, Chap. 7),[16] and Norton's dome isn't a representation of any actual building (Norton 2003, 2008). In addition to representations like four-sex population models and Norton's dome, which were never intended to be representations of an actual system, there are also models that were originally introduced as representational models but then turned out to be targetless. Maxwell's model of the ether is a case in point. While Maxwell didn't think that the ether consisted of wheels and pulleys, he thought that this model represented a real substance, namely ether. By modern lights there is no ether, and so Maxwell's model turned out to be targetless. Irrespective of whether models are designed from the outset as targetless models or whether they turn out to be such in the course of history, these models seem to be representations of some sort, and they have representational content. The question is: how can that be and what is their representational content? An account of representation has to be able to explain how such models work without being representations of a target, and one would hope that such an account would also provide an explication of the role such models play in scientific practice.[17] This is our fourth condition of adequacy, which we call the *Targetless Representations Condition*.

The question of what counts as a targetless model is subtle, and much depends on how the target is conceptualised. Consider the example of phlogiston. According to contemporary chemistry, phlogiston does not exist. So it would seem natural to deem a model constructed on the basis of Priestley's chemistry, which would involve phlogiston, as a targetless model. But whether it really is a targetless model depends on what exactly the model is taken to represent. If the model represents phlogiston specifically and aims to portray the various properties of the substance, then the model is targetless if phlogiston does not exist. One could, however, interpret a

---

[15] This is not to say that one cannot build the *aim* of accurate representation into the analysis of the concept of epistemic representation, or that the question of in virtue of what something is an epistemic representation need be entirely divorced from the question of in virtue of what something is an accurate epistemic representation. See Bolinska's (2013, 2016) and Poznic's (2018) for discussions of this point.

[16] Weisberg also talks about three-sex population models that were originally introduced without reference to a target system. Interestingly, it was later discovered that there are some ant populations which can be plausibly described as consisting of three sexes. The philosophical point remains: for *n*-sex population models, with $n > 3$, there are no actual target systems, for all we know.

[17] See Luczak's (2017) for a related discussion.

"Priestley style" model as a model of combustion, and thus interpreted the model is not targetless because the process of combustion obviously exists. In fact a phlogiston model and an oxygen model then have the same target, and the difference between them is only that the former is an inaccurate representation while the latter is an accurate representation (at least by our current lights). The model then would simply be a misrepresentation of combustion because it attempts to explain combustion in terms of an entity that does not exist. Thus understood, a phlogiston model is a misrepresentation in much the same way in which Ptolemy's model is a misrepresentation: it represents a real phenomenon, namely planetary motion, as being the result of epicycles, which, however, don't exist. This makes Ptolemy's model a misrepresentation, but not targetless model. So deciding whether a model qualifies as targetless requires careful attention to what exactly one takes the target to be.

In making this decision, two further issues have to be taken into account. First, there are different kinds of target systems. Familiar cases of targets involve models of *particular* target systems, for instance when we model the solar system. However, models can also represent *kinds* of systems, for instance when we model the hydrogen atom. Plausibly some models represent *general features* of the world, without necessarily targeting any particular system, which is what happens, for instance, when a model in elementary particle physics represents symmetry breaking without representing that process in a particular setting. The latter two cases raise ontological questions (do kinds exist? do general features exist?), and whether a model is classified as targetless will depend on how these questions are answered.

Second, earlier we said that a target system is a selected part or aspect of the world. But what does that mean? Here and throughout the book, we adopt a minimal form of realism. We assume that target systems exist independently of human observers, and that they are how they are irrespective of what anybody thinks about them. That is, we assume that the targets of representation exist independently of the representation. This is a presupposition not everybody shares. Constructivists (and other metaphysical antirealists) believe that phenomena do not exist independently of their representations because representations *constitute* the phenomena they represent.[18] This might lead some to draw the conclusion that all models are targetless because their targets don't exist. An assessment of the constructivist programme is beyond the scope of this book. It is worth observing, though, that many of the discussions to follow are by no means obsolete from a constructivist perspective. What, in the realist idiom, is conceptualised as the representation of an object in the world by a model would, in the constructivist idiom, be conceptualised as the relation between a model and *another* representation, or an object constituted by another representation. This is because even from a constructivist perspective, models and their targets are not identical, and the fact that targets are representationally constituted would not obliterate the differences between a target and its model.

---

[18] This view is expounded, for instance, by Lynch and Woolgar (1990). Curran (2018) and Giere (1994) offer critical reflections on the view.

1.4 The Application of Mathematics

Many models are mathematized, and their mathematical aspects are crucial to their cognitive and representational functions. Although relatively simple, the Newtonian model mentioned in the Introduction involved calculus, and many models in contemporary science rely on structures from highly abstract realms of mathematics. This forces us to reconsider a time-honoured philosophical puzzle: the applicability of mathematics in the empirical sciences. Even though the problem can be traced back at least to Plato's *Timaeus*, its canonical modern expression is due to Wigner, who famously remarked that "the enormous usefulness of mathematics in the natural sciences is something bordering on the mysterious and that there is no explanation for it" (1960, p. 2). One need not go as far as seeing the applicability of mathematics as an inexplicable mystery, but the question remains: how does mathematics hook onto the world?

The recent discussion of this problem has taken place in a body of literature that grew out of the philosophy of mathematics, and there has been little contact with the literature on modelling.[19] This is a regrettable state of affairs. The question of how a mathematized model represents its target subsumes the question of how mathematics applies to a physical system. So rather than separating the question of scientific representation from the problem of the applicability of mathematics, and dealing with them in separate discussions, they should be seen as the two sides of the same coin and be dealt with in tandem. For this reason, our fifth and final condition of adequacy is that an account of representation has to explain how mathematics is applied to the physical world in the context of mathematised representations. We call this the *Applicability of Mathematics Condition*.

1.5 The Carriers of Representations

When stating the different problems of representation, we spoke of the object that is doing the representing as the "carrier" of a representation. The question then is: what are carriers? Some carriers are ordinary material objects. The Phillips-Newlyn machine, which we will discuss in some detail later in the book, is an economic model consisting of pipes and reservoirs with water flowing through them (see Sect. 8.1); sausages of plasticine are used as models of proteins (de Chadarevian 2004);

---

[19] For a review of this literature see Shapiro's (1997, Chap. 8). Bueno and Colyvan (2011) offer a discussion of the applicability of mathematics that touches on the question of scientific representation. We discuss their approach in Chaps. 4 and 9, as well as in our (2017). It's worth noting that there is a distinction between the questions of whether (or not) mathematics performs an *explanatory* or a *representational* role in science (Saatsi 2011b). We're focussed on the latter here.

and oval shaped blocks of wood serve as models of ships (Leggett 2013). We call these kinds of representations *material* carriers.[20]

However, not all carriers are mannerly material objects. Some, like the Newtonian model of planetary motion, are, to use Ian Hacking's (1983, p. 216) memorable phrase, things that "you hold in your head rather than your hands". But what kind of things do we hold in our heads? We use the term *non-material* carrier to describe these representations. The negative characterisation is deliberate. At this point we want to remain non-committal and leave it open what kind of things such carriers are, or, indeed, whether they are things at all. To come to grips with these issues is what we call the *Problem of Carriers*.[21]

The Problem of Carriers can be divided into two subproblems. The first is the *Problem of Ontology*. In most basic terms, ontology is the study of what there is.[22] Ontological problems concern the question of whether or not something exists, and, if it does, what it is and what sort of stuff it is made of. Asking an ontological question in the context of epistemic representation amounts to asking whether carriers are objects, and, if the answer to this question is affirmative, asking what the nature and properties of these objects are. Possible answers, which we will touch on in the chapters to follow, include that they are mathematical structures, equations, abstract objects, fictional objects, possible worlds, or non-existent Meinongian objects. Some authors reject an understanding of models as "things" and push a programme that can be summed up in the slogan "modelling without models" (Levy 2015). If there are no models, then there is no problem concerning their ontology, and accordingly proponents of the modelling without models programme reject the Problem of Ontology as a non-issue.

The second subproblem of the Problem of Carriers concerns all questions that arise in the practice of handling carriers in the context of scientific modelling. What problems these are depends both on how the other problems of representation are answered and on what carriers one is willing to accommodate in one's theory of representation. Among the issues that arise in scientific practice when scientists use representations the following five are particularly important (Frigg 2010a, pp. 256–57). We refer to this (somewhat loosely circumscribed) set of issues as the *Problem of Handling* because these issues concern questions of how carriers are handled in the scientific process.[23]

---

[20] Alternative labels are "concrete model" (Thomson-Jones 2012, p. 761; Weisberg 2013, p. 24) and "physical model" (O'Connor and Weatherall 2016 p. 615).

[21] Non-material models raise a number of philosophical issues and have therefore attracted much attention in debates about representation. It is therefore worth pointing out that material models also raise interesting questions. Rosenblueth and Wiener (1945) discuss the criteria for choosing an object as a model. Ankeny and Leonelli (2011) discuss issues that arise when using organisms as models. The contributors to Klein's (2001) discuss representation in the laboratory.

[22] The term "ontology" derives from the ancient Greek "ὄντος" (ontos), meaning "that which is", and "λογος" (logos), meaning "logical discourse". So ontology literally means something like the study of being.

[23] We note here that we have previously used the phrase "Problem of Ontology" to refer to the wider problem, which here we call "Problem of Carriers" (Frigg and Nguyen 2016a, Sect. 1; 2017a,

1.5 The Carriers of Representations

The first issue is identity: every account of carriers has to provide *identity conditions for carriers*. This is because carriers are often presented in different ways by different authors. And variations of this kind are not merely a matter of personal style. As Vorms points out (2011, 2012), models can be presented under different "formats", meaning that they are described using different conceptual resources and, if the model is mathematized, also different formalisms. The format of a model is important because how scientists use a model and what inferences they draw from it crucially depends on the format.[24] The intuition here is that alternative descriptions nevertheless describe the same model. But when is it the case that two descriptions describe the same model and when are the resulting models different? There is nothing special about models in that regard: many carriers can, in principle, be described in different ways and so the question of identity equally arises for many carriers.

The second issue is *property attribution*. It is a common practice in science to attribute properties to carriers. We say that the period of the ideal pendulum is one second, that the isolated population grows monotonically, and that the goods changed hands between two agents for £5. This does not seem to sit well with the fact that the relevant carriers often are, as we have just seen, things that we hold in our heads rather than our hands. What sense does it make to say that the pendulum has a period of one second or that the population grows monotonically if these carriers are non-material, and therefore are not the kinds of things to which we usually attribute physical features like oscillation periods or monotonic population growth? Unless one wants to countenance the unappealing proposition that much of scientific discourse is incoherent, one has to understand the semantics of such statements.

The third issue is related to the second and concerns *comparative statements*. Scientists not only attribute properties to carriers, they also compare them with parts of reality, or with another carriers, or both. This happens when we say things like "John does not behave like the rational agent in the model" and "the shell model of the nucleus has a more complex internal structure than the liquid drop model". How can we compare a non-material carrier to a concrete target, and, possibly even more problematic, how can we compare two non-material carriers with each other?

The fourth issue concerns *truth in a carrier*. There is right and wrong in a discourse about carriers. It's true that the motion of planets in Newton's model is confined to a plane and it's false that those planets move in parabolic orbits. On what basis are claims about a model system qualified as true or false, in particular if the claims concern issues about which the original model-description remains silent? It was neither part of the original specification of Newton's model that the trajectories are confined to a plane; nor was it stated that trajectories aren't parabolic. Yet both claims are true in the model. To come to grips with this issue we need an account of

---

Sect. 1). As we will see in the chapters to come, much of the discussion concerning the "ontology" of models actually addresses the Problem of Handling, rather than the underlying ontological nature of scientific models. For this reason it turned out to be useful to separate the problem of ontology and the problem of handling for the extended discussion in this book.

[24] See also Knuuttila's (2011, 2017), where she emphasises the importance of understanding the ways in which models are given to scientists.

carriers that first, explains what it means for a claim about a carrier to be true or false, and second, draws the line between true and false statements in the right place (for instance, an account on which all statements about a carriers come out false would be unacceptable).

The fifth and final issue within the Problem of Handling is the *epistemology of carriers*. Scientists investigate carriers and find out what properties they have. Newton found out that his model-planets move in elliptical orbits. Discovering what is true in carriers is an essential part of scientific investigation; if truths about carriers were forever concealed from scientists, then representations would have little if any scientific value. But how do scientists find out about carriers' properties and how do they justify their claims?

This completes our discussion of the five problems that an account of representation has to solve, and of the five conditions of adequacy on acceptable solutions. We summarise the problems and the conditions, along with their interrelations, in Fig. 1.2. These problems and conditions will accompany us throughout the book. We will look at different accounts of representation and for each account ask how it answers these problems, and whether the answers meet the conditions. In this way our problems and conditions provide a unified framework to first discuss different accounts (Chap. 2 through to Chap. 7), and to then develop our own proposal (Chaps. 8 and 9).

1.6 Chapter Summary

In this section we provide a concise summary of the five problems that a philosophical account of scientific representation has to solve, and of the five conditions of adequacy that an answer to these problems has to meet, which we have introduced, motivated, and justified in Sect. 1.1 through to Sect. 1.5.

The *target system* (or *target*) of a representation is the part or aspect of the world that is represented by a representation. The object that performs the representational function is the representation's *carrier*. Sir Isaac is the target of his portrait, and the canvass covered with pigments is the carrier of the portrait.

Models, theories, pictures, photographs, maps, diagrams, charts, drawings, statues, and movies are examples of representations that afford their users epistemic access to features of their targets (at least potentially). One can study these representations and in doing so discover, or at least attempt to discover, features of the representations' targets. We call such representations *epistemic representations*, and the process by which a user of the representation extracts information about a target from a representation *surrogative reasoning*. Epistemic representations are the focus of this book.

Every analysis has to begin by circumscribing its subject matter. In our case this means that every account of representation has to specify which of a range of epistemic representations it aims to address. This gives rise to our first problem, the *Representational Demarcation Problem*: the question whether different sorts of

1.6 Chapter Summary

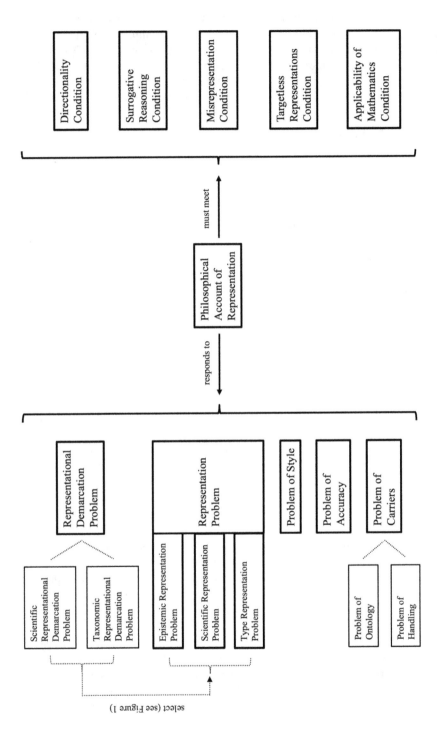

Fig. 1.2 Summary of the problems and conditions of adequacy. The use of connected boxes rather than dotted lines for the Representation Problem indicates that the three problems attached to it are different versions of the Representation Problem rather than subproblems

representations have to be distinguished and, if so, what these sorts are. How one responds to this problem has important methodological implications. The problem can be divided into two sub-problems. The *Scientific Representational Demarcation Problem* is to answer the question whether scientific representations differ from other kinds of epistemic representations as far as their representational characteristics are concerned, and, if so, to pinpoint wherein the difference lies. The *Taxonomic Representational Demarcation Problem* is to answer the question whether there are different types of representations, and, if so, identify the differences in their representational characteristics. Do we have to distinguish, say, between diagrams, pictures, and models, and then analyse them separately? Both sub-problems concern the question whether there are differences as regards the *representational characteristics* of different representations. Representations may serve different social functions and be associated with different modes of production, but differences of this kind are irrelevant from the point of view of a theory of representation. The two demarcation problems are independent in that one can give a positive answer to one and negative answer to the other.

The second problem is the *Representation Problem*: what turns something into a representation of something else? Depending on what answers one gives to the two demarcation problems, one will address different versions of the Representation Problem. The different versions and their dependence on the answers to the demarcation problems are summarised in the matrix in Fig. 1.1 (in Sect. 1.1). Those who give negative answers to both demarcation problems can directly address the *Epistemic Representation Problem* (*ER-Problem*), which consists in filling the blank in the biconditional "X is an epistemic representation of T iff ___", where "X" stands for the carrier of the representation and "T" for the target system. We call this biconditional the *ER-Scheme*. Those who give a positive answer to the *Scientific Representational Demarcation Problem* will strive to offer a solution to the *Scientific Representation Problem* (*SR-Problem*) which consists in filling the blank in the biconditional "X is a scientific representation of T iff ___", where the biconditional is the *SR-Scheme*. Those who respond affirmatively to the *Taxonomic Representational Demarcation Problem* will have to say what types of representation they recognise, and each type Y will then require its own analysis, which amounts to filling blank in "X is a scientific/epistemic representation of type Y of T iff ___". We call this the *Type Representation Problem* (*TR-Problem*), and the corresponding biconditional is the *TR-Scheme*. Whether one addresses the scheme for "scientific" or "epistemic" depends on one's answer to *Scientific Representational Demarcation Problem*. Someone who buys into scientific demarcation will solve the TR-Problem for "scientific representation"; someone who does not will solve it for "epistemic representation".

The first two problems are the most difficult problems on our list, but being clear about the aims of one's analysis is important to avoid complications down the line. Things will be more straightforward from now on. The third problem is to respond to what we call the *Problem of Style*: what styles are there and how can they be characterised? This problem is motivated by the fact that there can a great variety of representational strategies even within a certain type of representation. Assume you demarcate in both ways and identify scientific models as one type. Then it is still the case that not all models represent their targets in the same way. For instance, some models are analogical representations of their targets while others are idealisations. An account of representation will have to identify different ways in which models can represent their targets and gain an understanding of these ways by distinguishing, and analysing, different representational styles.

Some representations are accurate; others are not. So the fourth problem is the *Problem of Accuracy*: under what conditions is a representation accurate? An account of representation has to say what accuracy is, and provide standards thereof.

The fifth and last problem is the *Problem of Carriers*. Previously we said that carriers are the objects that are doing the representing. Broadly speaking, the Problem of Carriers is to understand what objects carriers are and how scientists engage with them in scientific practice. The problem can be broken down in two subproblems. The *Problem of Ontology* is the question: are carriers objects, and, if so, what kind of objects are they? The *Problem of Handling* is the question: how are carriers handled in the process of using a representation? This is particularly pertinent with representations of the kind discussed in the Introduction, where the carrier is not a physical object.

Not any response to these problems is a good response. But how do we distinguish between good and bad responses? We submit that responses should be judged against five conditions of adequacy, and a good account has to meet all of them. It is coincidence that there are five problems and also five conditions, and problems are not "paired up" with questions.

The first condition of adequacy is the *Directionality Condition*. Representation is directed. X represents T, but typically not vice versa. For example, the portrait of Newton represents Newton, but Newton does not represent his portrait. The first condition says that any acceptable account of representation has to regard representation as directional in this way and identify the root of representation's directionality.

As we noted at the beginning of this section, the representations that we are interested in are epistemic representations. These are representations that allow for surrogative reasoning: one can study a representation and in doing so discover, or at least attempt to discover, features of the things that the representation stands for. An account of representation must explain how surrogative reasoning is possible. This is the *Surrogative Reasoning Condition*, our second condition of adequacy.

Some representations are misrepresentations. The third condition of adequacy, the *Misrepresentation Condition*, requires that an account of representation make room for misrepresentation. That is, an account must not conflate misrepresentation and non-representation. Misrepresentations are wrong representations, but they are representations nevertheless.

The fourth condition is the *Targetless Representations Condition*. It is a fact of scientific practice that there are models with no target systems at all. Maxwell's model of the ether is a case in point. There is no ether and yet there is a model – seemingly – representing the ether. An account of representation must explain what happens in such cases.

At least in scientific contexts, many representations are highly mathematised. Elaborate mathematical formalisms are part of the representational apparatus, and target systems are, or seem to be, represented as having certain mathematical properties. This brings back the well-known problem of the applicability of mathematics: how does mathematics latch onto reality? The fifth and last condition requires an account of representation to explain how the mathematical apparatus used in a carrier attaches to the physical world. This is the *Applicability of Mathematics Condition*.

We now have our five problems and five conditions in front of us. They are summarised visually in Fig. 1.2 (at the end of Sect. 1.5). This sets the agenda. We will now see what responses current accounts of representation offer to these problems and examine these responses against our conditions of adequacy.

Chapter 2
General Griceanism and Stipulative Fiat

General Griceanism is the proposal that all types of representation can be explained in a unified way as deriving from a more fundamental kind of representations, namely mental states. The view thus holds that there is no problem of scientific representation, or of any other form of epistemic representation like images, diagrams, or graphs. The only form of representation that requires a sustained philosophical investigation is mental states, and all other representations derive from these through an act of Stipulative Fiat.

We begin the chapter by discussing how General Griceanism deals with the demarcation problems and then state its response to the ER-Problem (Sect. 2.1). In the previous chapter we formulated a number of problems and conditions that every account of scientific representation has to meet, and so we continue our discussion by investigating whether, and if so how, General Griceanism fits the bill (Sect. 2.2). We conclude the chapter with an enquiry into the relation between General Griceanism and Stipulative Fiat, and argue that they are in fact two separate doctrines (Sect. 2.3).

2.1 Demarcation and the ER-Problem

Callender and Cohen (2006) submit that the entire debate over scientific representation got started on the wrong foot. They claim that scientific representation is not different from "artistic, linguistic, and culinary representation" and in fact "there is no special problem about scientific representation" (ibid. p. 67). Underlying this claim is a position Callender and Cohen call "General Griceanism" (GG):

> The General Gricean holds that, among the many sorts of representational entities (cars, cakes, equations, etc.), the representational status of most of them is derivative from the representational status of a privileged core of representations. The advertised benefit of this General Gricean approach to representation is that we won't need separate theories to

account for artistic, linguistic, [...] and culinary representation; instead, the General Gricean proposes that all these types of representation can be explained (in a unified way) as deriving from some more fundamental sorts of representations, which are typically taken to be mental states. (Of course, this view requires an independently constituted theory of representation for the fundamental entities.) (ibid. p. 70)

GG then comes with a practical prescription about how to proceed with the analysis of a representation:

the General Gricean view consists of two stages. First, it explains the representational powers of derivative representations in terms of those of fundamental representations; second, it offers some other story to explain representation for the fundamental bearers of content. (ibid., p. 73)

Importantly, according to Callender and Cohen, of these two stages only the second stage requires serious philosophical work. The first involves only "a relatively trivial trade of one problem for another" (ibid.). The second stage is a task for the philosophy of mind because the fundamental form of representation is mental representation.[1]

Scientific representation is a derivative kind of representation (ibid., p. 71, 75) and hence falls under the first stage of the above recipe. It is reduced to mental representation by an act of stipulation:

Can the salt shaker on the dinner table represent Madagascar? Of course it can, so long as you stipulate that the former represents the latter. [...] Can your left hand represent the Platonic form of beauty? Of course, so long as you stipulate that the former represents the latter. [...] On the story we are telling, then, virtually anything can be stipulated to be a representational vehicle for the representation of virtually anything [...]; the representational powers of mental states are so wide-ranging that they can bring about other representational relations between arbitrary relata by dint of mere stipulation. The upshot is that, once one has paid the admittedly hefty one-time fee of supplying a metaphysics of representation for mental states, further instances of representation become extremely cheap. (ibid., pp. 73–74)

Explaining any form of representation other than mental representation is then a triviality: all it takes is an act of "stipulative fiat" (ibid., p. 75). Once we understand how mental states represent, there is no problem left to answer. Scientific representation only requires "an act of stipulation to connect representational vehicle with representational target" (ibid., p. 79), and this is all that needs to be said about how objects in science come to represent.

Callender and Cohen explicitly discuss what we call the Scientific Representational Demarcation Problem and submit that they "are not optimistic about solving this problem" (ibid., p. 83). We take it from their discussion that optimism is also not required because, from the point of view of GG, the answer to the problem is negative. Once scientific representation is declared to be on par with artistic, linguistic, and other forms of representation, scientific demarcation becomes obsolete: scientific representations are just one kind of epistemic representation among many

---

[1] Ducheyne seems to holds a similar position when he declares that "[s]cientific models are essentially a subset of mental representations" (2008, p. 120).

2.1 Demarcation and the ER-Problem

others and there is no need to demarcate between them, at least with respect to their representational status. Callender and Cohen don't explicitly discuss the Taxonomic Representational Demarcation Problem, but given their insistence that there are only two kinds of representation – mental states and derivative representations – it would seem natural not to demarcate taxonomically either: scientific models are just one kind of scientific representation among many.[2] So they will end up answering the ER-problem and not distinguish between scientific and other kinds of non-fundamental representations.

The view that scientific representation requires nothing over and above an act of stipulation supplies Callender and Cohen's answer to the ER-problem, which we call Stipulative Fiat for obvious reasons:

Stipulative Fiat: a carrier X is an epistemic representation of a target system T iff a user stipulates that X represents T.

On this view, scientific representations are cheap to come by. The question therefore arises why scientists spend a lot of time constructing and studying complex models if they might just as well take a salt shaker and turn it into a representation of, say, a Bose-Einstein condensate by an act of fiat. Why bother taking the time to use the Newtonian model of the planet's orbit if one could just as well stipulate that it's represented by a wine glass? Callender and Cohen admit that there are useful and not so useful representations, and that salt shakers belong in the latter group. However, they insist that this has nothing to do with representation: "the questions about the utility of these representational vehicles are questions about the pragmatics of things that are representational vehicles, not questions about their representational status *per se*" (ibid., p. 75). So whilst the usefulness of a representation might turn on other relations, such as isomorphism, similarity, or inference generation, these features serve pragmatic purposes without establishing representation relations:

> We are not denying that isomorphism, similarity, and inference generation may relate representational vehicle and representational target in many cases of scientific (and other non-natural) representation. We claim that these conditions do not constitute the representational relation, and hence are not necessary features of representation. However, we allow that there are important roles for these conditions —viz., they may serve as pragmatic aids to the recognition of a representational relation that is constituted by other means. (ibid., p. 78)

So, in sum, if epistemic representation "is constituted in terms of a stipulation, together with an underlying theory of representation for mental states, isomorphism, similarity, and inference generation are all idle wheels" (ibid.).

---

[2] However, it is worth noting that there is nothing in GG than prescribes that the reduction of derivative representation to fundamental reduction be done the same way across different instances of scientific and epistemic representation. The difference between GG and Stipulative Fiat is discussed below.

2.2 Taking Stock

Let's now see how Stipulative Fiat fares with respect to the other problems and conditions of adequacy formulated in the previous chapter.

There are two ways in which to think about the Problem of Style through the lens of Stipulative Fiat. One would either focus on the representation relation, and therefore it would seem like there is only one way of turning something into an epistemic representation, and therefore there is only one style of (non-derivative) representation, namely the "stipulation style". Or alternatively, as is suggested by Callender and Cohen, one could address the Problem of Style through an investigation of the different relations – isomorphism, similarity, or inference generation, for example – which serve non-representational, pragmatic purposes:

> there will remain a role for considerations about isomorphism, similarity, and inference generation after all. Namely, these considerations (and possibly others) may contribute to an anthropology of the use of scientific representations by providing a taxonomy of the sorts of pragmatically guided heuristics scientists bring to bear on their choices between representational vehicles. (2006, p. 78)

This could then lead to the recognition of different styles like the isomorphism-style or the similarity-style. Whilst this is an option for Callender and Cohen, they say little about what styles are. Likewise, they do not address the Problem of Accuracy and do not comment on standards of accuracy that would pertain to each of these styles.

Callender and Cohen have also little to say about the Problem of Carriers. This is hardly surprising given that they explicitly recognise objects as diverse as salt shakers, cakes, and equations as possible carriers. Stipulative Fiat as a position does not require any restrictions and is compatible with various different kinds of carriers.

Does Stipulative Fiat meet our conditions of adequacy? The most troubling condition is the Surrogative Reasoning Condition, which we will discuss in the next section. Callender and Cohen sharply distinguish between the ER-Problem and the question of what makes a representation accurate (ibid. p. 69), and they take themselves to be addressing the former. As such, their account does allow for the possibility of misrepresentation because nothing in a stipulation requires that the representation be accurate (although a question remains about what exactly it would mean for a stipulation to be inaccurate).

Callender and Cohen do not explicitly discuss how Stipulative Fiat meets the Targetless Representations Condition, but since nothing stops a model user from stipulating that X represents a non-existent target, Stipulative Fiat seems to have an obvious response to this condition. An analysis of what such an act of stipulation involves could proceed in the same way as an analysis of bearer-less names or descriptions.

Stipulative Fiat also naturally meets the Directionality Condition. A model is stipulated to be a representation of a target, and the act of stipulation is clearly directional: a user's stipulation that X represents T does not imply that T also represents X. Callender and Cohen do not comment on the applicability of mathematics,

2.3 General Griceanism and Stipulative Fiat

The most troubling question we are faced with when assessing this account is the relation between GG and Stipulative Fiat, and, as we will see, this has significant knock-on effects for how Callender and Cohen's discussion fares with respect to the Surrogative Reasoning Condition. Callender and Cohen do not comment in detail on the relationship between the two, but that they mention both in the same breath would suggest that they regard them as one and the same doctrine, or at least as the two sides of the same coin.[3] This is not so. Stipulative Fiat is just one way of fleshing out GG, which only requires that there be *some* explanation of how derivative representations relate to fundamental representations. GG does not require that this explanation be of a particular kind, much less that it consist in nothing but an act of stipulation. Toon (2010, pp. 77–78; 2012b, p. 244) rightly points out that it is one thing to claim that the representation relation between a carrier and a target only holds in virtue of some other, more fundamental kind of representation; it is another thing to claim that the relation is one of pure stipulation. Scientific representation can, in principle, be "reduced" to fundamental representation in many different ways (some of which we will encounter later in this book). Conversely, the failure of Stipulate Fiat does not entail that we must reject GG: one can uphold the idea that an appeal to the intentions of model users is a crucial element in an account of scientific representation even if one rejects Stipulative Fiat (again, this idea will reappear later in the book).

Let us now examine Stipulative Fiat. Callender and Cohen emphasise that anything can be a representation of anything else: "as a reflex of its generality, the explanatory strategy we are now considering places almost no substantive constraints on the sorts of things that can be representational relata" (2006, p. 73). This is correct. Things that function as models don't belong a distinctive ontological category, and it would be a mistake to think that some objects are, intrinsically, representations and other are not. This point has been made by others too,[4] and, as we shall see, it is a cornerstone of several alternative accounts of representation.

---

[3] They also mention a position they call "Specific Griceanism" (2006, p. 72) from the philosophy of language, but they don't discuss how Specific Griceanism works in the context of scientific representation. One thing worth highlighting here is that according to Specific Griceanism the meaning of linguistic representations is determined by the intentions of the speaker to activate certain beliefs in the audience. Whilst Callender and Cohen do not endorse Specific Griceanism as a solution to the ER-Problem, it does highlight the possibility that a successful answer to the ER-Problem may need to include reference to an audience. We return to this briefly in Sect. 3.2.

[4] See, for instance, Frigg's (2010b, p. 99), Frisch's (2014, p. 3028), Giere's (2010, p. 269), Suárez's (2004, p. 773), Swoyer's (1991, p. 452), Teller's (2001, p. 397), van Fraassen's (2008, p. 23), and Wartofsky's (1979, p. xx).

But even if anything can, in principle, be a representation of anything else, it doesn't follow that a mere act of stipulation suffices to turn an object into a representation of a target. Furthermore, it doesn't follow that an object elevated to the status of a representation by an act of fiat represents its target in a way that can appropriately be characterised as an instance of epistemic representation. We discuss both concerns in reverse order.

As Toon (2010, pp. 78–79) points out, the term "representation" is flexible enough to support different uses. And, at least in some sense, we might accept that Callender and Cohen are correct to claim that the salt shaker represents Madagascar in virtue of someone stipulating that this is the case. However, the ER-problem does not concern representation in this broad sense. The crucial question is not whether the salt shaker represents Madagascar, but rather whether it represents in a way similar to the way in which the Watson-Crick model represents DNA, the Newtonian model represents the planet's orbit, and a portrait represents its subject: namely in a way that allows users of the representation to extract claims about the target from the representation. And it doesn't seem like it does. Stipulative Fiat fails to meet the Surrogative Reasoning Condition because it fails to provide an account of how reasoning about the salt shaker would allow for reasoning about Madagascar.

Even if we admit that Stipulative Fiat establishes that models *denote* their targets (and as we will see soon, there is a question about this), denotation is not sufficient for epistemic representation. To use Toon's analogy from pictorial representation, the word "Napoleon" and Jacques-Louis David's portrait of Napoleon both serve to denote the French general. But this does not imply that they represent him in the same way.[5] Morrison (2015, p. 127) agrees and points out that Callender and Cohen's "solution overlooks what is significant about scientific representation, namely the larger role that representation plays in generating scientific knowledge". Bueno and French (2011, pp. 871–74) gesture in the same direction when they point to Peirce's distinction between icon, index and symbol and dismiss Callender and Cohen's view on grounds that they cannot explain the obvious differences between different kinds of representations. Boesch (2017) points out that Stipulative Fiat fails to account for how scientific representations licence inferences about their target systems. Finally, as Gelfert notes:

> Models can surprise us, open up unforeseen lines of inquiry, and lead to novel insights about their targets, all of which suggests that they enjoy considerable autonomy. Mathematical models, in particular, are imbued with a considerable internal structure and dynamics, which renders them partially independent from the intentions of their users. None of this is easily captured by Callender's and Cohen's account, which accords them only an auxiliary role as vehicles of pre-existing intentions and beliefs on the part of their users. (2016, p. 33)

---

[5] Callender and Cohen seem sensitive to this when they claim that "the Gricean story we are telling allows for two distinct but related sorts of representation [… o]n the one hand, there is representation of things[… o]n the other hand there is representation of facts" (2006, p. 74). The former is simply denotation; the latter is a more complex representational relationship akin to the epistemic representation relation we are interested in. The problem is that Stipulative Fiat fails to distinguish between these types of representation, which it both sees as established by stipulation.

Supporters of Stipulative Fiat could try to mitigate the force of this objection in two ways. First, they could appeal to additional facts about the carrier, as well as its relation to other items, in order to account for surrogative reasoning. For instance, the salt shaker being to the right of the pepper mill might allow us to infer that Madagascar is to the east of Mozambique. Moves of this sort, however, invoke (at least tacitly) a specifiable relation between features of the model and features of the target (similarity, isomorphism, or otherwise), and an invocation of this kind goes beyond mere stipulation. Second, the last quotation from Callender and Cohen in Sect. 2.1 suggests that they might want to relegate surrogative reasoning into the realm of pragmatics and deny that it is part of the relation properly called representation. This, however, in effect amounts to a removal of the Surrogative Reasoning Condition from the desiderata of an account of scientific representation, and we have argued in Chap. 1 that surrogative reasoning is one of the hallmarks of scientific representation (and even if it were "merely" pragmatics, we still would want an account of how it works).

Let us now turn to our first point, that a mere act of stipulation is insufficient to turn X into a representation of T. We take our cue from a parallel discussion in the philosophy of language, where it has been pointed out that it is not clear that stipulation is sufficient to establish denotation (which is weaker relation than epistemic representation). A position similar to Stipulative Fiat faces what is known as the *Humpty Dumpty Problem*, named in reference to Lewis Carroll's discussion of Humpty using the word "glory" to mean "a nice knockdown argument" (MacKay 1968; cf. Donnellan 1968).[6] If stipulation is all that matters, then as long as Humpty simply stipulates that "glory" means "a nice knockdown argument", then it does so. And this doesn't seem to be the case. Even if the utterance "glory" *could* mean "a nice knockdown argument" – if, for example, Humpty was speaking a different language – in the case in question it doesn't, irrespective of Humpty's stipulation. In the contemporary philosophy of language the discussion of this problem focuses more on the denotation of demonstratives rather than proper names, and work in that field focuses on propping up existing accounts so as to ensure that a speaker's intentions successfully establish the denotation of demonstratives uttered by the speaker (see Michaelson and Reimer 2019, Sect. 3.2 and the references therein) Whatever the success of these endeavours, their mere existence shows that successfully establishing denotation requires moving beyond a bare appeal to stipulation, or brute intention. But if a brute appeal to intentions fails in the case of demonstratives – the sorts of terms that such an account would most readily be applicable

---

[6] The Humpty Dumpty Problem concerns the general question whether or not a speaker's utterance *means* whatever the speaker intends the utterance to mean. A closely related problem is whether a speaker's bare intention to *denote* suffices to establish that a proper name denotes whatever the speaker intends it to denote.

to – then we find it difficult to see how Stipulative Fiat will establish a representational relationship between models and their targets.[7]

It now pays that we have separated GG from Stipulative Fiat. Even though Stipulative Fiat does not provide an adequate answer to the ER-problem, one can still uphold GG. As Callender and Cohen note, all that it requires is that there is a privileged class of representations,[8] and that other types of representations owe their representational capacities to their relationship with these privileged ones. So philosophers need an account of how members of this privileged class of representations represent, and how derivative representations, which includes scientific models, relate to members of this class. This is a plausible position and, put in these terms, Callender and Cohen's discussion serves to encourage philosophers of science interested in the relationship between scientific models and their targets to pay attention to how analogous debates are carried out in other areas of philosophy. Understanding the ways in which others have investigated how representation works in the context of linguistic or pictorial representation, and how this connects to the mental states and intentions of speakers, artists, and audiences, has the potential to further our understanding of how scientific representation works (in this sense we share Butterfield's (2020) endorsement of this aspect of Callender and Cohen's account).

Moreover, when stated like this, many recent contributors to the debate on scientific representation can be seen as falling under the umbrella of GG. As we will see below, the more developed versions of the similarity (Chap. 3) and isomorphism (Chap. 4) accounts of scientific representation make explicit reference to the intentions and purposes of model users, even if their earlier iterations did not. And so do the accounts discussed in the later chapters of this book, where the intentions of model users (in a more complicated manner than that suggested by Stipulative Fiat) are invoked to establish epistemic representation. Indeed we can grant that scientific representation requires intentionality, but it doesn't follow that intentionality is all that is required to establish it. So what we take issue with when it comes to GG is not the claim that scientific representation is a derivative kind of representation – many would accept this. But rather, we object to the claim that the relationship between scientific representation and the more primitive kinds of representation is a trivial act of stipulation.

---

[7] Moreover, this whole discussion presupposes that an intention-based account of denotation is the correct one. This is controversial: see Michaelson and Reimer's (2019) for an overview of discussions of denotation in the philosophy of language. If this is not the correct way to think about denotation, then Stipulative Fiat will fail to get off the ground.

[8] They take them to be mental states but are open to the suggestion that they might be something else (2006, p. 82).

Chapter 3
The Similarity View

Moving on from the Gricean account, we now turn to the similarity view of scientific representation.[1] Similarity and representation initially appear to be two closely related concepts, and invoking the former to ground the latter has a philosophical lineage stretching back at least as far as Plato's *The Republic*. In its most basic guise the similarity view of representation says that something represents something else by being similar to it. A photograph represents its subject matter by being similar to it; a statue represents its subject by being similar to it; and a painting represents by being similar to what it portrays. In the context of a discussion of modelling, the similarity view asserts that scientific models represent their targets in virtue of being similar to them.[2]

The similarity view offers an elegant answer to the problem of surrogative reasoning. Similarities between a model and its target can be exploited to carry over insights gained in the model to the target. If the similarity between a model and its target is based on shared properties, then a property found in the model would also have to be present in the target; and if the similarity holds between properties themselves, then the target would have to instantiate properties similar to the model.

We begin the chapter by discussing how the similarity view deals with the demarcation problems (Sect. 3.1), and then turn to how it deals with epistemic representation (Sect. 3.2). We then discuss the similarity view's take on the issues of accuracy

---

[1] This conception is also known as the *resemblance view of representation*. "Resemblance" seems to be more commonly used in aesthetics (in particular in discussions about pictorial representation), while "similarity" is the term of choice in philosophy of science.

[2] Sometimes the point is made by categorising models as icons in Peirce's sense, which Kralemann and Lattmann claim are "characterized by having a representational quality on the basis of a similarity relation between themselves and their objects" (2013, p. 3398). See also Ambrosio's (2014) and Gallegos's (2019).

and style (Sect. 3.3), and we end by considering its stance on the Problem of Carriers (Sect. 3.4).

3.1 The Demarcation Problems

The similarity view has often been presented as accounting for a wide range of representations. Pictures of different kinds (including paintings, drawings, and photographs), three-dimensional representations (such as statues, mock-ups, and architectural models), and, of course, scientific models, are said to represent by being similar to their respective targets.[3] This suggests that its proponents take the similarity view be an account of epistemic representation that covers scientific and non-scientific representations alike. If so, they have to give a negative answer to the Scientific Representational Demarcation Problem, which would require a separation of scientific from other epistemic representations.

Things get more interesting when we turn to the Taxonomic Representational Demarcation Problem. This issue is brought into focus through Weisberg's (2007b) distinction between direct and indirect representation. Weisberg submits that modelling is different from other kinds of scientific theorising, and that this difference matters to how one analyses the working of their respective representations. Models, Weisberg argues, are distinctive in that they are *indirect representations.* When a scientist builds a model, she constructs a model object, which then becomes the focus of study. She analyses the object, draws conclusions about it, and then assesses the object's relation to the intended target. At this stage similarity becomes crucial because if the model is deemed "sufficiently similar to the world, then the analysis of the model is also, indirectly, an analysis of the properties of the real-world phenomenon", which is why "modeling involves indirect representation and analysis of real-world phenomena via the mediation of models" (ibid., pp. 209–10). The important point here is that the model is seen as an object that has properties, that has an internal dynamic, and that behaves in a certain way. This point is obvious when the model is a material object like the ones we have mentioned in the Introduction, where, for instance, a system of pipes and reservoirs is used to model an economy or, to take one of Weisberg's own examples, a pond equipped with pumps is used to model the San Francisco Bay area. But the point equally holds for non-material models like Newton's model of the solar system, where imaginary spherical planets are the objects of study. The label "indirect" is owed to the fact that models are often specified through model-descriptions. Newton's model is introduced through a set of sentences like the ones at the beginning of the Introduction. On the face of it these descriptions might look like plain descriptions of a target, but they are not. They

---

[3] In this chapter we focus on scientific models. Abell (2009), Blumson (2014), and Lopes (2004) provide discussions of similarity in the context of visual representation.

3.1 The Demarcation Problems

Fig. 3.1a Indirect representation

Fig. 3.1b Direct representation

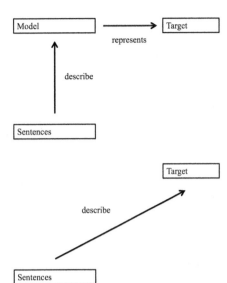

describe a model, and the model in turn represents the target.[4] Indirect representation is illustrated in Fig. 3.1a.

Indirect representation contrasts with *direct representation*, which amounts to representing targets without the mediation of a secondary system.[5] A description of a target system with English sentences is a direct description. Weisberg provides the example of Darwin's geological work, which offers descriptions of atolls and coral islands, and he sets out to formulate a theory about their origin and development (ibid., p. 227). Darwin does so not by constructing models that then represent atolls, but rather by offering a direct description of them. But direct representation is not limited to linguistic description. Weisberg also mentions Mendeleev's periodic table as an example of a direct representation because it describes the properties of chemical elements by referring to them directly and not by specifying models of them (ibid., p. 215). Direct representation is illustrated in Fig. 3.1b.

This amounts to a response to the Taxonomic Representational Demarcation Problem: in contrast with other kinds of representations that represent directly,

---

[4] For further discussion of modelling as indirect representation see Giere's (1999; 2006, Chap. 4), Godfrey-Smith's (2006), and Knuuttila's (2017). The idea is also discussed in Chaps. 4 and 6. The specification of a model can also rely on visual rather than verbal means of communication (Giere 1996).

[5] Weisberg speaks of "direct abstract representation" to indicate that theories offer general, or "abstract", representations. We drop this qualification here because we are not concerned with that aspect of theories. Liu draws a distinction similar to Weisberg's when he contrasts "purely symbolic" representations which are "conventional", with scientific models, which are "epistemic vehicles" (2015a, p. 41; cf. 2015b, p. 287).

models are indirect representations.[6] So the similarity view rejects scientific demarcation whilst upholding taxonomic demarcation. This places the similarity view in the bottom left matrix element of the matrix in Fig. 1.1. This means that the similarity view has to address the TR-Problem for "epistemic representation", which is to fill the blank in the scheme "X is an epistemic type Y representation of T iff ___" for every type Y. Since the similarity view recognises two types of epistemic representation, direct and indirect representation, this amounts to completing two schemes, namely "X is a direct epistemic representation of T iff ___" and "X is an indirect epistemic representation of T iff ___". Given that Weisberg associates direct representation with linguistic descriptions, completing the scheme for direct epistemic representation is largely a problem for the philosophy of language. For this reason, proponents of the similarity view can set this problem aside and focussed on the scheme for indirect epistemic representation, which is where similarity plays a crucial role. We call the problem of filling the blank in "X is an indirect epistemic representation of T iff ___"; the *Indirect Epistemic Representation Problem*.

This way of dealing with the problems stems from the fact that the distinction between direct and indirect representation is not specific to scientific representation, and thus demarcating taxonomically in terms of directness cuts across the science/non-science distinction (and thus does not force a defender of the similarity view to demarcate scientifically). Weisberg introduces the distinction in the context of non-material models where a linguistic description is used to specify the model, and the model in turn represents the target by entering into a similarity relation with it. The core of this view is that a carrier X represents by being similar to the target T. This analysis is then extended to material models in his (2013). Moreover, as we have seen at the beginning of this section, paintings, statues, and other objects can also be seen as representing by being similar to their targets. Thus, they are also classified as indirect representations. We note that the term "indirect" is less apt in these cases than it is when discussing non-material models, because, unlike non-material models, images and statues need not be specified linguistically. But, as we have just seen, the core of the similarity view's treatment of indirect epistemic representation is that representation is rooted in a similarity relation between an object and its subject, and hence the terminological stretch of referring to them as "indirect" representations is harmless. From this point of view, images and statues offer indirect representations while verbal descriptions offer direct representations of their respective subject matters. William Turner's painting *The London Bridge Coach At Cheapside* indirectly represents a part of London. It is an object, like a model, that

---

[6] In their review of Weisberg's (2013), O'Connor and Weatherall (2016) suggest that Weisberg should demarcate even further: even "models" is too broad a unit of analysis. Their claim is that given the important differences between models from across various scientific disciplines, an analysis that groups them together is too general for the result to be informative. As we will see in the remainder of this book, this level of generality need not result in uniformativeness, at least when the issue of representation is at stake. Furthermore, as we discuss in Sect. 5.3, accounts of epistemic representation can be given which, whilst stated abstractly and thus covering epistemic representations in general, need to be concretized in different ways in different cases.

can be seen as representing its target through being similar to it. Peter Ackroyd's book *London: The Biography*, by contrast, offers a direct description of parts of London in the same way in which Darwin's writings on atoll formation describe atolls.

3.2 The Indirect Epistemic Representation Problem

Appeal to similarity in the context of epistemic representation leaves open whether similarity is offered as an answer to the Indirect Epistemic Representation Problem, the Problem of Style, or the Problem of Accuracy. Proponents of the similarity account typically offer little guidance on this issue. So we examine each option in turn and ask whether similarity offers a viable answer in the sense of meeting the conditions of adequacy introduced in Chap. 1.

Understood as response to the Indirect Epistemic Representation Problem, a similarity view of representation amounts to the following:[7]

Similarity 1: a carrier X is an indirect epistemic representation of a target T iff X and T are similar.

A well-known objection to this account is that similarity has the wrong logical properties. Goodman (1976, pp. 4–5) submits that similarity is symmetric and reflexive, yet representation isn't. If object A is similar to object B, then B is similar to A. But if A represents B, then B need not (and in fact in most cases does not) represent A: the Newtonian model represents the orbiting planet, but the planet does not represent the Newtonian model. And everything is similar to itself, but most things do not represent themselves.[8] So this account does not meet our first condition of adequacy for an account of epistemic representation because it does not provide a direction to representation.

Yaghmaie (2012) argues that this conclusion – along with the Directionality Condition itself – is wrong: epistemic representation is symmetric and reflexive.[9] His examples are drawn from mathematical physics, and he presents a detailed case study of a mutual representation relation between quantum field theory and statistical mechanics. His case raises interesting questions, but even if one grants that Yaghmaie has identified a case where epistemic representation goes both ways it does not follow that this is a *general* feature of (indirect) epistemic representation: at best it shows that epistemic representation is not asymmetric. The photograph in Jane's passport represents Jane; but Jane does not represent her passport photograph. The same holds true for myriads of other representations. Goodman is cor-

---

[7] Positions like Similarity 1 have been formulated (but not endorsed) by Goodman (1976, p. 3) and Suárez (2003, p. 227).

[8] Problems similar to these also arise in connection with other logical properties. For a discussion of transitivity see Frigg's (2003, p. 31) and Suárez's (2003, pp. 232–33).

[9] Yaghmaie discusses logical properties in the context of the isomorphism view of representation (to which we turn in the next chapter). The point, however, equally applies to the similarity view.

rect in pointing out that *typically* representation is not symmetrical and reflexive: a target T does not represent X just because X represents T, and X does not represent itself.

A reply diametrically opposed to Yaghmaie's emerges from Weisberg's (2013), which builds on Tversky's (1977, 1978) account of similarity. Weisberg accepts that representation is not symmetric, but disputes that similarity fails on this count. Using a gradual notion of similarity (i.e. one that allows for statements like "A is similar to B to degree d"), Tversky found that subjects in empirical studies judged that North Korea was more similar to China than China was to North Korea (Tversky and Gati 1978). Likewise, discussing the characters in a Polanski movie Poznic (2016b, Sect. 4.2) points out that the similarity relation between a baby and the father need not be symmetric.

So allowing degrees into one's notion of similarity makes room for cases of non-symmetry.[10] This raises the question of how to analyse similarity. We discuss this thorny issue in some detail in the next section. For now we concede the point and grant that similarity need not always be symmetrical. However, this does not solve Goodman's problem with reflexivity (as we will see, on Weisberg's notion of similarity everything is maximally similar to itself); nor does it, as we will see now, solve other problems of the similarity account.

Regardless of how the issue of logical properties is resolved, there is another serious problem: similarity is too inclusive a concept to account for epistemic representation. In many cases neither one of a pair of similar objects represents the other. Two copies of the same book are similar but they don't represent each other. And this problem won't go away even if similarity turns out to be non-symmetric. That North Korea is similar to China (to some degree) does not imply that North Korea represents China, and that China is not similar to North Korea to the same degree does not alter this conclusion. This point has been brought home in now-classical thought experiment due to Putnam (1981, pp. 1–3).[11] An ant is crawling on a patch of sand and leaves a trace that happens to resemble Winston Churchill. Has the ant produced a picture of Churchill? Putnam's answer is that it didn't because the ant has never seen Churchill and had no intention to produce an image of him. Although *someone else* might see the trace as a depiction of Churchill, the trace itself does not represent Churchill. This, Putnam concludes, shows that "[s]imilarity […] to the features of Winston Churchill is not sufficient to make something represent or refer to Churchill" (ibid., p. 1). And what is true of the trace and Churchill is true of every other pair of similar items: similarity on its own does not establish representation.

There is also a more general issue concerning similarity: it is too easy to come by. Without constraints on what counts as similar, any two things can be considered similar to any degree (Aronson et al. 1995, p. 21). Later in the chapter we will see

---

[10] Metric-based notions of similarity will introduce a notion of degree whilst remaining committed to symmetry. Tversky's account explicitly does not utilize a metric.

[11] Black (1973, p. 104) discusses a very similar thought experiment.

3.2 The Indirect Epistemic Representation Problem

reasons for holding this view. For now, note that if correct, this, combined with Similarity 1, has the unfortunate consequence that anything represents anything else because any two objects are similar in some respect. Similarity is just too inclusive to account for representation.

An obvious response to this problem is to delineate a set of relevant respects and degrees to which X and T have to be similar. This suggestion has been made explicitly by Giere (1988, p. 81) who suggests that carriers and their targets have to be similar in relevant respects and to certain degrees. This idea can be moulded into the following definition:

Similarity 2: a carrier X is an indirect epistemic representation of a target T iff X and T are similar in relevant respects and to the relevant degrees.

On this definition one is free to choose one's respects and degrees so that unwanted similarities drop out of the picture. While this solves the last problem, it leaves the others untouched: similarity in relevant respects and to the relevant degrees is reflexive (and symmetrical, depending on one's notion of similarity); and presumably the ant's trace in the sand is still similar to Churchill in the relevant respects and degrees but without representing Churchill. Moreover, Similarity 2 introduces three new problems.

First, a misrepresentation is one that portrays its target as having properties that are not similar in the relevant respects and to the relevant degrees to the true properties of the target. But then, on Similarity 2, a misrepresentation is not a representation at all. In Chap. 1, we argued that the possibility of misrepresentation is a condition of adequacy for any acceptable account of representation and so we submit that misrepresentation should not be conflated with non-representation.[12]

Second, similarity in relevant respects and to the relevant degrees does not guarantee that X represents the right target. As Suárez points out (2003, pp. 233–34), even regimented similarity can obtain with no corresponding representation. He offers the example of a friend who dresses up as Pope Innocent X (and does so perfectly). The friend then resembles Velázquez's portrait of the pope (at least in as far as the pope himself resembled the portrait). Yet the portrait, despite being similar to the friend, represents the pope and not the friend. In cases like these, which Suárez calls "mistargeting", a model represents one target rather than another, despite the fact that both targets are relevantly similar to the model. Like in the case of Putnam's ant, the root cause of the problem is that the similarity is accidental. In the case of the ant, the accident occurs at the representation-end of the relation, whereas in the case of the friend's dressing up the accidental similarity occurs at the target-end. Both cases demonstrate that Similarity 2 cannot rule out accidental representation.

---

[12] Ducheyne (2008) offers a variant of a similarity account which explicitly takes the *success* of the hypothesized similarity between a model and its target to be a necessary condition on the model representing the target. This approach also fails to account for the possibility of misrepresentation. Since he has more recently offered an alternative account of scientific representation, which we discuss in Sect. 5.4, we do not address his previous proposal here.

Third, there simply may be nothing to be similar to because some representations represent no actual object (Goodman 1976, p. 26). Some paintings represent elves and dragons, and some models represent phlogiston and the ether. None of these exist. As Toon (2012b, pp. 246–47) points out, this is a problem in particular for the similarity view: models without objects cannot represent what they seem to represent because in order for two things to be similar to each other both have to exist. If there is no ether, then an ether model cannot be similar to the ether.

It would seem that at least the second problem could be solved by adding the requirement that X denote T.[13] Amending the previous definition accordingly yields:

Similarity 3: a carrier X is an indirect epistemic representation of a target T iff X and T are similar in relevant respects and to the relevant degrees, and X denotes T.

This account would also solve the problem with reflexivity (and symmetry), because denotation is directional in a way similarity is not. Unfortunately Similarity 3 still suffers from the first and the third problems. It would still lead to the conflation of misrepresentations with non-representations because the first conjunct (similar in the relevant respects) would still be false. A non-existent system cannot be denoted and so we have to conclude that a model of, say, a four-sex population represents nothing. This seems unfortunate because there is a clear sense in which models without targets are about something. Maxwell's work on the ether provides a detailed and intelligible account of a number of properties of the ether, and these properties are highlighted in the model. If ether existed then Similarity 3 could explain why these were important by appealing to them as being relevant for the similarity between an ether model and its target. But since ether does not exist, no such explanation can be offered. Yet the model has representational content, which Similarity 3 cannot capture.

A different version of the similarity view sets aside the moves made in Similarity 3 and tries to improve on Similarity 2. The crucial move then is to take the very act of *asserting* a specific similarity between a model and a target as constitutive of indirect epistemic representation.

Similarity 4: a carrier X is an indirect epistemic representation of a target system T iff a theoretical hypothesis H asserts that X and T are similar in relevant respects and to the relevant degrees.

This comes close to the view Giere advocated in *Explaining Science* (1988, p. 81).[14] This version of the similarity view avoids problems with misrepresentation because, being a hypothesis, there is no expectation that the assertions made in H are true. If they are, then the representation is accurate (or the representation is accurate to the extent that they hold). If they are not, then the representation is a misrepresentation. It resolves the problem of mistargeting because hypotheses iden-

---

[13] A proposal along those lines was mentioned (but, again, not endorsed) by Goodman (1976, pp. 5–6).
[14] Something like this view can also be found in Cartwright's (1999a, pp. 192–93; 1999b, pp. 261), where she appeals to a "loose notion of resemblance".

3.2 The Indirect Epistemic Representation Problem 39

tify targets before asserting similarities with X (that is, the task of picking the right target is now placed in the court of the hypothesis and is no longer expected to be determined by the similarity relation). Finally it also resolves the issue with directionality because H can be understood as introducing a directionality that is not present in the similarity relation. However, it fails to resolve the problem with representation without a target. If there is no T, then it's not clear that a hypothesis can be asserted about T.

Let us set the issue of non-existent targets aside for the moment and have a closer look at the notion of representation proposed in Similarity 4. A crucial point remains understated in Similarity 4. Hypotheses don't assert themselves. Hypotheses are put forward by those who work with representations, in the case of models by scientists. So the crucial ingredient – users – is left implicit in Similarity 4.

In two more recent publications Giere made explicit the fact that "scientists are intentional agents with goals and purposes" (2004, p. 743) and proposed to build this insight into an account of indirect epistemic representation. This involves adopting an agent-based notion of indirect epistemic representation that focuses on "the activity of representing" (ibid., p. 743). Analysing indirect epistemic representation in these terms amounts to analysing schemes like "S uses X to represent W for purposes P" (ibid.), or in more detail: "Agents (1) intend; (2) to use model, M; (3) to represent a part of the world W; (4) for purposes, P. So agents specify which similarities are intended and for what purpose" (2010, p. 274). This conception of representation was already proposed half century earlier by Apostel when he put forward the following analysis of scientific representation: "Let then $R(S, P, M, T)$ indicate the main variables of the modelling relationship. The subject S takes, in view of the purpose P, the entity M as a model for the prototype T" (Apostel 1961, p. 4).[15] In our own terminology, Apostel's and Giere's proposals come down to:

Similarity 5: a carrier X is an indirect epistemic representation of a target system T iff there is an agent A who uses X to represent a target system T by proposing a theoretical hypothesis H specifying a similarity (in relevant respects and to the relevant degrees) between X and T for purpose P.

This definition inherits from Similarity 4 the resolutions of the issues concerning directionality, misrepresentation, and mistargeting; and for the sake of argument we assume that the problem with non-existent targets can be resolved in one way or another.

Assigning an essential role to the purposes and actions of scientists in the definition of epistemic representation marks a significant departure from previous similarity-based accounts. Suárez, drawing on van Fraassen's (2002) and Putnam's (2002), defines "naturalistic" accounts of representation as ones in which "whether

---

[15] A very similar idea is also described by Wartofsky, who takes representation to be triadic relation between a subject, a model, and target (1979, p. 6). This "intentional turn" is now widely accepted. A dissenting position is Rusanen and Lappi's who emphasise that "the semantics of models as scientific representations should be based on the mind-independent model-world relation" (2012, p. 317).

or not representation obtains depends on facts about the world and does not in any way answer to the personal purposes, views or interests of enquirers" (2003, pp. 226–27). By building the purposes of carrier users directly into an answer to the Indirect Epistemic Representation Problem, Similarity 5 is explicitly not a naturalistic account (in contrast, for example, to Similarity 1). As we noted in Chap. 1, what grounds epistemic representation does not need to depend solely on features of X and T, and therefore a response to the Indirect Epistemic Representation Problem does not have to be naturalistic in Suárez's sense. Moreover, as we saw in Chap. 2, General Griceanism actively recommends that the purposes, views, or interests of enquirers should be built into the conditions on epistemic representation. We agree that this is the right move, and, as we will discuss later, many of the more developed answers to versions of the epistemic representation problem in other intellectual traditions also share this feature.

Does this suggest that Similarity 5 is a successful solution to the Indirect Epistemic Representation Problem? Unfortunately not. A closer look at Similarity 5 reveals that the role of similarity has shifted. As far as offering a solution to the Indirect Epistemic Representation Problem is concerned, all the heavy lifting in Similarity 5 is done by the appeal to agents, and similarity has in fact become an idle wheel. Giere implicitly admits this when he writes:

> How do scientists use models to represent aspects of the world? What is it about models that makes it possible to use them in this way? One way, perhaps the most important way, *but probably not the only way*, is by exploiting similarities between a model and that aspect of the world it is being used to represent. Note that I am not saying that the model itself represents an aspect of the world because it is similar to that aspect. There is no such representational relationship. [footnote omitted] Anything is similar to anything else in countless respects, but not anything represents anything else. *It is not the model that is doing the representing; it is the scientist using the model who is doing the representing.* (2004, p. 747, emphasis added)

But if similarity is not the only way in which a carrier can be used as a representation, and if it is the use by a scientist that turns a carrier into a representation (rather than any mind-independent relationship the carrier bears to the target), then similarity has become otiose in a reply to the Indirect Epistemic Representation Problem. A scientist could invoke any relation between X and T, and X would still represent T. Being similar in the relevant respects to the relevant degrees now plays the role either of a representational style, or of a normative criterion for accurate representation, rather than of a grounding of representation. We assess in the next section whether similarity offers a cogent reply to the Problem of Style and the Problem of Accuracy.

The shift from thinking about carriers as representing to users as performing representational actions invites a further question: at least in the context of scientific representation, what does it mean for a scientist to perform a representational action? What is it about the scientist invoking similarity, or another relation, between a model and a target that makes the action representational? Boesch (2019a) draws

3.2 The Indirect Epistemic Representation Problem

from the literature on the philosophy of action to attempt to answer this question.[16] He utilises Anscombe's (2000, 2005) account of intentional action. According to her, what makes something an intentional action is that it can be described by a series of means-end ordered descriptions. To use Boesch's example, when someone pours herself a cup of coffee, the action can be described as "tipping a pot of coffee", "getting a refill", and "rewarding oneself for finishing a section" (Boesch 2019a, p. 2311). These descriptions describe the same action, and are ordered in such a way that earlier descriptions are the means for the ends that feature in latter ones. Boesch then argues that something is a "scientific, representational action" (ibid., p. 2312) when the descriptions are licenced by scientific practice, and where a description of a scientist's interaction with a model stands as an earlier description towards the final description which is some scientific aim such as explanation, predication, or theorizing (thereby making the action scientific). In order to make the action representational, according to Boesch, it must be the case that these actions are described in such a way that an "earlier description involves interaction with a vehicle and a later description involves the (potential) achievement of some end about the target" (ibid. p. 2317). Boesch's discussion indicates how shifting from thinking about representation as a relation to a practice might be further explicated, and it seems to offer a strategy to engage with the wider philosophical literature in doing so.

But as an answer to Indirect Epistemic Representation question (or in Boesch's case the SR-question, since he demarcates scientific representational actions from non-scientific ones) the account remains opaque. The idea that an action has a "denotational form" (ibid.), as Boesch calls it, when it can be described by a sequence of descriptions where an earlier description describes interacting with a model and a latter investigating a target seems to just restate the problem. In Boesch's framework the SR-question becomes: in virtue of what can an action be described in these two different ways? Saying that they can amounts to saying that they are representational actions without providing an account of what makes them so. But without an answer to the latter, the account remains, at the very least, incomplete.

Mäki (2011, p. 59) suggested an extension of Similarity 5, which he explicitly describes as "as a (more or less radical) modification of" Giere's.[17] Mäki adds two conditions to Giere's: the agent uses the model to address an audience E and offers a commentary C (ibid., p. 57). The role of the commentary is to specify the nature of the similarity. This is needed because "representation does not require that all parts of the model resemble the target in all or just any arbitrary respects, or that the issue of resemblance legitimately arises in regard to all parts. The relevant model parts and the relevant respects and degrees of resemblance must be delimited"

---

[16] It's worth noting that Boesch's discussion isn't limited to a similarity account, but is supposed to supplement various different non-naturalistic accounts with an account of representational action. See his (2019b) for a discussion of how different accounts of scientific representation invoke the intentional actions of model users in, according to Boesch, compatible ways.

[17] See also his (2009).

(ibid.). What these relevant respects and degrees of resemblance are depends on the purposes of the scientific representation in question. These are not determined "in the model", as it were, but are pragmatic elements. From this it transpires that in effect C plays the same role as that played by theoretical hypotheses in Giere's account. Certain aspects of X are chosen as those relevant to the representational relationship between X and T.

The addition of an audience, however, is problematic. While models are often shared publicly, this does not seem to be a necessary condition for the representational use of a model.[18] There is nothing that precludes a lone scientist from coining a model and using it representationally.[19] That some models are easier to grasp and therefore serve as more effective tools to drive home a point in certain public settings is an undisputable fact, but one that has no bearing on a model's status as a representation. Furthermore, if the audience is part of representation it follows that the representation changes whenever the audience changes. Reiss argues, rightly in our view, this is an undesirable consequence because models don't change when a teacher uses the same model in two different classrooms (2013, p. 108). But maybe changing classrooms is too small a change. The representation certainly would change if the model was moved from a classroom to a rocket science laboratory, which would involve a change in the *kind* of audience that the model has (namely from students to rocket scientists). Reiss agrees, but notes that even then Mäki's account misidentifies the reason why the representation changes. What drives the change is not a difference in audience but a difference in purpose (ibid.): what makes the representation different is that rocket scientists pursue different aims in using a model and draw different inferences from the model than students.

The conclusion we draw from this discussion is that similarity does not offer a viable answer to the Indirect Epistemic Representation Problem.

3.3 Accuracy and Style

Accounting for the possibility of misrepresentation resulted in a shift of the division of labour for the more developed similarity-based accounts. Rather than being the relation that grounds epistemic representation, similarity should be considered as answering the Problem of Accuracy or the Problem of Style (or both). The former is motivated by the observation that a proposed similarity between X and T could be

---

[18] We say "shared publically" because one could of course stipulate that a scientist is his or her own audience, and thus an audience is necessary for representation. This would be a spurious response. The question at hand here is whether an audience beyond the agent who is doing the representing is required.

[19] Although see Osbeck and Nersessian's (2006) for a discussion of the idea that representation is an aspect of distributed cognition. See also Boesch's (2017), which draws on Suárez's (2010, p. 99) for an argument that a wider scientific community is needed to fix the rules used according to which a model can be used for surrogative reasoning.

3.3 Accuracy and Style

wrong, and hence if the user's proposal does hold (and X and T are in fact similar in the specified way) then X is an accurate representation of T. The latter transpires from the basic fact that a judgment of accuracy in fact presupposes a choice of respects in which X and T are claimed to be similar. Simply proposing that they are similar in some unspecified respect is vacuous. But delineating relevant properties could potentially provide an answer to the Problem of Style. For example, if X and T are proposed to be similar with respect to their causal structure, then we might have a style of causal modelling; if X and T are proposed to be similar with respect to structural properties, then we might have a style of structural modelling; and so on and so forth. So the idea is that if X representing T involves the claim that X and T are similar in a certain respect, the respect chosen specifies the style of the representation; and if X and T are in fact similar in that respect (and to the specified degree), then X accurately represents T within that style.

In this section we investigate both options. But before delving into the details, let us briefly step back and reflect on possible constraints on viable answers. Taking his cue from Lopes' (2004) discussion of pictures, Downes (2009, pp. 421–22) proposes two constraints on allowable notions of similarity. The first, which he calls the *independence challenge*, requires that a user must be able to specify the relevant similarity *before* engaging in a comparison between X and T. Similarities that are recognisable only with hindsight don't provide a viable foundation of a representation. We agree with this requirement. In fact, it is closely related to the Surrogative Reasoning Condition: a carrier can generate novel hypotheses only if (at least some of the) similarity claims are not known only ex post facto.

Downes' second constraint, the *diversity constraint*, is the requirement that the relevant notion of similarity has to be identical in all kinds of representation and across all representational styles. So all carriers must bear the same similarity relations to their targets. Whatever its merits in the case of pictorial representation, this observation does not hold water in the case of scientific representation. Teller (2001a, p. 402) insisted – rightly, in our view – that there need not be a substantive sense of similarity uniting all scientific representations. A proponent of the similarity view is free to propose different kinds of similarity for different representations and is under no obligation to also show that they are special cases of some overarching conception of similarity.

We now turn to the Problem of Style. A first step in the direction of an understanding of styles is the explicit analysis of the notion of similarity. Unfortunately the philosophical literature contains surprisingly little by way of an explicit discussion of what it means for something to be similar to something else. In many cases similarity is taken to be primitive, possible worlds semantics being a prime example. The problem is then compounded by the fact that the focus is on comparative overall similarity rather than on similarity in respects and to degrees.[20] Where similarity in respects is discussed explicitly, the standard way of cashing out what it means for an object to be similar to another object is to require that they co-

---

[20] For a critical discussion of overall similarity see Morreau's (2010).

instantiate properties. This is the idea that Quine (1969, pp. 117–18) and Goodman (1972, p. 443) had in mind in their influential critiques of the notion. They note that if all that is required for two things to be similar is that they co-instantiate *some* property, then everything is similar to everything else, since any two objects have at least one property in common.

The issue of similarity seems to have attracted more attention in psychology. In fact, the psychological literature provides formal accounts to capture similarity directly in fully worked out accounts. The two most prominent suggestions are the *geometric* and *contrast* accounts (see Decock and Douven's (2011) for a recent discussion). The former, associated with Shepard's (1980), assigns objects a place in a multidimensional space based on values assigned to their properties. This space is then equipped with a metric and the degree of similarity between two objects is a function of the distance between the points representing the two objects in that space. This account is based on the strong assumptions that values can be assigned to all features relevant to similarity judgments, which is deemed unrealistic. Although similarity with respect to quantitative features might be the sort of thing that could be assigned a value, it's not clear that this is the case for all instances of similarity. This problem is supposed to be overcome in Tversky's *contrast account* (1977).

This account defines a gradated notion of similarity based on a weighted comparison of properties. Weisberg (2012; 2013, Chap. 8) has recently introduced this account into the philosophy of science where it serves as the starting point for his so-called *weighted feature matching account of model world-relations*. This account is our primary interest here. The account introduces a set Δ of relevant features. Let then $\Delta_X \subseteq \Delta$ be the set of features from Δ that are instantiated by the carrier X; likewise let Δ_T be the set of features from Δ instantiated by the target system. Furthermore let f be a weighting function assigning a real number to every subset of Δ. The simplest version of a weighting function is one that assigns to each set the number of features in the set, but weights can be more complex, for instance by assigning more important features higher values. The level of similarity between X and T is then given by the following equation (Weisberg 2012, p. 788):[21]

$$S(X,T) = \theta f(\Delta_X \cap \Delta_T) - \alpha f(\Delta_X - \Delta_T) - \beta f(\Delta_T - \Delta_X),$$

where $\Delta_X \cap \Delta_T$ denotes the set of features from Δ that X and T have in common; $\Delta_X - \Delta_T$ denotes the features (from Δ) that X has but T does not; $\Delta_T - \Delta_X$ denotes the features (from Δ) that T has but X does not; and the α, β, and θ are further weights, which can in principle take any value, capturing the relevant importance of shared and unshared properties. This equation provides a "similarity score that can be used in comparative judgments of similarity" (ibid., p. 788). Thus X and T are similar (according to S) to the extent that they share features from Δ (appropriately weighted by f and θ), penalised by the extent to which they differ with respect to those

---

[21] We have used our own notation to retain consistency with other sections.

3.3 Accuracy and Style

features (appropiately weighted by f, α, and β). In the above formulation the similarity score S can in principle vary between any two values (depending on the choice of the weighting function and the value of the further weights). One can then use standard mathematical techniques to renormalize S so that it takes values in the unit interval [0, 1] (these technical moves need not occupy us here and we refer the reader to Weisberg 2013, Chap. 8, for details).[22]

The obvious question at this point is how the various details in the account can be filled. First in line is the specification of a feature set Δ. Weisberg is explicit that there are no general rules to rely on and that "the elements of Δ come from a combination of context, conceptualization of the target, and theoretical goals of the scientist" (2013, p. 149). Likewise, the weighting function and the values of further weighting parameters depend on the goals of the investigation, the context, and the theoretical framework in which scientists operate.[23]

Irrespective of these choices, the similarity score S has a number of interesting features. First, it is asymmetrical for $\alpha \neq \beta$, which makes room for the possibility of X being similar to T to a different degree than T is similar to X. So S provides the non-symmetrical notion of similarity mentioned earlier in this chapter. Second, S has a property called *maximality*: everything is maximally similar to itself and every other non-identical object is equally or less similar. Formally: $S(A, A) \geq S(A, B)$ for all objects A and B (ibid., p. 154).

What does this account contribute to a response to the Problem of Style? The answer, we think, is that it has heuristic value but does not provide a substantive account. The idea is that different styles are associated with different choices of the kinds of features collected in Δ (what features do we focus on?), the weight given by f to features (what is the relative importance of features?) and the value of the parameters (how significant are disagreements between the features of X and T?). Different styles of representation could be grouped together in terms of how these choices are made. However, when thought of in these terms, stylistic questions stand outside the proposed framework. Questions concerning what the stylistic groupings are, even if answered in terms of the choices referenced above, have to be answered outside the account. The account offers a framework in which questions can be asked, but it does not itself provide answers and hence no classification of representational styles emerges from it.

Some will say that this is old news. Goodman denounced similarity as "a pretender, an impostor, a quack" (1972, p. 437) not least because he thought that it merely put a label to something unknown without analysing it. And even some

---

[22] In passing we note that this account could be seen as a quantitative version of Hesse's (1963) theory of analogy. Properties that X and T share are the *positive analogy* and ones they don't share are the *negative analogy*. Weisberg's formula then gives a comparative measure of positive and negative analogy.

[23] Weisberg further divides the elements of Δ into *attributes* and *mechanisms*. The former are the "the properties and patterns of a system" while the latter are the "underlying mechanism[s] that generates these properties" (ibid., p. 145). This distinction is helpful in the application to concrete cases, but for the purpose of our conceptual discussion it can be set aside.

proponents of the similarity view have insisted that no general characterisation of similarity was possible. Thus Teller submits that "[t]here can be no general account of similarity, but there is also no need for a general account because the details of any case will provide the information which will establish just what should count as relevant similarity in that case" (2001a, b, p. 402). This amounts to nothing less than the admission that no analysis of similarity (or even different kinds of similarity) is possible at a general level and that we have to deal with each case in its own right.

Assume now, for the sake of argument, that the stylistic issues have been resolved and full specifications of relevant features and their relative weights are available. It would then seem plausible to say that $S(X, T)$ provides a degree of accuracy. This reading is supported by the fact that Weisberg paraphrases the role of $S(X, T)$ as providing "standards of fidelity" (2013, p. 147).[24]

As we have seen above, $S(X, T)$ is maximal if X is a perfect replica of T (with respect to the features in Δ), and the fewer features (in Δ) X and T share, the less accurate the representation becomes. This lack of accuracy is then reflected in a lower similarity score. This is plausible, at least for some kinds of epistemic representation, and Weisberg's account is indeed a step forward in the direction of quantifying accuracy.[25]

Weisberg's account is an elaborate version of the co-instantiation account of similarity. It improves significantly on simple versions, but it cannot overcome that account's basic limitations. Niiniluoto (1988, pp. 272–74) distinguishes between two different kinds of similarities: partial identity and likeness.[26] Assume X instantiates the relevant features $P_1, \ldots, P_n$ and T instantiates the relevant features $Q_1, \ldots, Q_n$. If these features are identical, i.e. if $P_i = Q_i$ for all $i = 1, \ldots, n$, then X and T are similar in the sense of being *partially identical* ("partially" because we are restricting our focus to relevant features). Partial identity contrasts with what Niiniluoto

---

[24] As Parker (2015) points out, it is unclear whether Weisberg's account offers standards of accuracy or an answer to what we call the Indirect Epistemic Representation Problem. Somewhat surprisingly, in his response to Parker's review Weisberg argues that it is supposed to do both. With respect to the Indirect Epistemic Representation Problem he claims that since "[w]ith no parameters set and no weighting function defined, the equation describes an infinite set of potential relations—at least one of which almost certainly holds between a model and a target. So it is correct to say that [the function S] is an account of the sort of relation that generally holds between models and their targets" (Weisberg 2015, p. 300). As we have seen in Sect. 3.2, this ubiquity is a vice rather than virtue and so it is remains unclear how this is supposed to be an answer to the Indirect Epistemic Representation Problem.

[25] This is not to say that increasing similarity will also entail increased representational utility. As mentioned in Sect. 1.3, Borges, and Carroll's parables involving of maps which are exact copies of their target terrain, and thus useless as maps, demonstrate that increasing similarity doesn't necessarily improve the representation. This is also discussed by Boesch (2019c).

[26] A similar distinction also appears in Hesse's discussion of analogies; see, for instance Hesse's (1963, pp. 66–67).

calls *likeness*. X and T are similar in the sense of likeness if the features are not identical but similar themselves: P_i is similar to Q_i for all $i = 1, ..., n$. So in likeness the similarity is located at the level of the features themselves. For example, a red post box and a red London bus are similar with respect to their colour, even if they do not instantiate the exact same shade of red. As Parker (2015, p. 273) notes, Weisberg's account (like all co-instantiation accounts) deals well with partial identity, but has no systematic place for likeness.

To deal with likeness Weisberg would in effect have to reduce likeness to partial identity by introducing "imprecise" features which encompass the P_i and the Q_i. Parker (ibid.) suggest that this can be done by introducing intervals in the feature set, for instance of the form "the value of feature [P] lies in the interval $[p - \varepsilon, p + \varepsilon]$" where ε is parameter specifying the precision of overlap. As an illustration she uses Weisberg's example of the San Francisco Bay model and submits that in order to account for the similarity between the model and the actual Bay with respect to their Froude number Weisberg has to claim something like: "The Bay model and the real Bay share the property of having a Froude number that is within 0.1 of the real Bay's number. It is more natural to say that the Bay model and the real Bay have *similar* Froude numbers—similar in the sense that their values differ by at most 0.1" (ibid.). In this way, whilst it may be the case that the model doesn't share the same exact property as the target they each have a property that lie in the same interval value, and thus share a less specific property.

In his response Weisberg accepts this and argues that he is trying to provide a reductive account of similarity that bottoms out in features shared and those not shared (2015, p. 302). But such interval-valued features have to be part of Δ in order for the formal account to capture them. This means that another important decision regarding whether or not X and T are similar occurs outside of the formal account itself. The inclusion criteria for what goes into Δ now not only have to delineate relevant features, but, at least for the quantitative ones, they also have to provide an interval defining when they qualify as similar. Furthermore, it remains unclear how to account for X and T to be alike with respect to their qualitative features. The similarity between genuinely qualitative properties cannot be accounted for in terms of numerical intervals. This is a particularly pressing problem for Weisberg, because he takes the ability to compare models and their targets with respect to their qualitative properties as a central desideratum for any account of similarity between the two (2013, p. 136). These sorts of considerations have led Khosrowi (2020) to suggest that the notion of "sharing features" should be taken to be analysed in a pluralist manner. Sharing a feature sometimes means sharing the exact same feature; sometimes it means to share features which are sufficiently quantitatively close to one another; and sometimes it means having features which are themselves "sufficiently similar". But if similarity is located between features themselves (rather than between particulars instantiating features), then one gives up on the idea of using the notion of sharing features to provide a reductive account of similarity, a conclusion that Khosrowi is prepared to endorse.

3.4 The Problem of Carriers

Another problem facing similarity-based approaches concerns their treatment of carriers. If carriers are supposed to be similar to their targets in the ways specified by theoretical hypotheses or commentaries, then they must be the *kind* of things that can be so similar.

As noted in Sect. 1.5, some models are homely physical objects. The Phillips–Newlyn model of an economy is a system of pipes and reservoirs; the Kendrew model of myoglobin is a plasticine sausage; and oval shaped blocks of wood are often used as models of ships. These are not isolated instances. The Army Corps of Engineers' model of the San Francisco Bay is a water basin equipped with pumps to simulate the action of tidal flows (Weisberg 2013); ball and stick models of molecules are made of metal or wood (Toon 2011); and model organisms in biology are animals like worms and mice (Ankeny and Leonelli 2011). For models of this kind similarity is straightforward (at least in principle) because they are of the same ontological kind as their respective targets: they are material objects.[27]

But many interesting models are not like this. Two perfect spheres with a homogeneous mass distribution which interact only with each other (the Newtonian model of a sun–earth system) or a single-species population isolated from its environment and reproducing at fixed rate at equidistant time steps (the logistic growth model of a population) are what Hacking aptly describes as something you hold in your head rather than your hands and we call non-material models. The question then is what kind of objects non-material models are. Giere submits that they are abstract objects: "Models in advanced sciences such as physics and biology should be abstract objects constructed in conformity with appropriate general principles and specific conditions" (2004, p. 745; cf. 1988, p. 81; 2010, p. 270).

The appeal to abstract entities brings a number of difficulties with it. The first is that the class of abstract objects is rather large. Numbers and other objects of pure mathematics, classes, propositions, concepts, the letter "A", and Dante's *Inferno* are abstract objects (Rosen 2020), and Hale (1988, pp. 86–87) lists no less than 12 different possible characterisations of abstract objects. At the very least this list shows that there is great variety in abstract objects and their characterisations. So classifying models as abstract objects adds little specificity to an account of what models are.

At this point the above discussion of direct vs. indirect representation becomes relevant. Recall the idea underlying indirect representation: model-descriptions specify model objects, and model objects represent target systems by being similar to them. In this sense, Giere could counter that he limits attention to those abstract objects that possess "all and only the characteristics specified in the principles"

---

[27] There is a possible complication here, if one considers the fact that there may be *multiple* concrete objects that might be thought to correspond to the same model. For example, we talk about "the" Phillips–Newlyn machine despite the fact that multiple such machines were made. Here a defender of the similarity account will need to distinguish between being similar to a token of a model, and being similar to the type of model. We set this complication aside here.

3.4 The Problem of Carriers

(2004, p. 745), where principles are general rules like Newton's laws of motion, possibly supplemented with initial and boundary conditions, that specify the models *qua* abstract objects. He further says that he takes "abstract entities to be human constructions" and that "abstract models are definitely not to be identified with linguistic entities such as words or equations" (ibid., p. 747). While this narrows down the choices somehow, it still leaves many options and ultimately the ontological status of models in the similarity account remains unclear.

Giere does not expand on this ontological issue for a reason: he dismisses the problem as one that philosophers of science can set aside without loss. He voices scepticism about the view that philosophers of science "need a deeper understanding of imaginative processes and of the objects produced by these process" (2009, p. 250) or that "we need say much more […] to get on with the job of investigating the functions of models in science" (ibid.).

We remain unconvinced about this scepticism, and even if one wants to remain silent about the Problem of Ontology, one will have to say something about the Problem of Handling.[28] Giere's quietism seems problematic because there is an obvious yet fundamental issue with abstract objects. No matter how the above issues are resolved (and irrespective of whether they are resolved at all), at the minimum it is clear that models are "abstract" in the sense that they have no spatiotemporal location. A number of authors have argued that this alone causes serious problems for the similarity account.[29] The similarity account demands that models can instantiate properties and relations, since this is a necessary condition on them being similar to their targets. In particular, it requires that models can instantiate the properties and relations mentioned in theoretical hypotheses or commentaries. But such properties and relations are typically *physical*. And if models have no spatiotemporal location, then they do not instantiate any such properties or relations. Thomson-Jones' example of the idealized pendulum model makes this clear. If the idealized pendulum is abstract then it is difficult to see how to make sense of the idea that it has a length, or a mass, or an oscillation period of any particular time.[30]

An alternative suggestion due to Teller is that we should instead say that whilst "concrete objects HAVE properties […] properties are PARTS of models" (2001a, p. 399, original capitalisation). It is not entirely clear what Teller means by this, but our guess is that he would regard models as bundles of properties. Target systems, as concrete objects, are the sorts of things that can instantiate properties delineated by theoretical hypotheses. Models, since they are abstract, cannot. But rather than being objects instantiating properties, a model can be seen as a bundle of properties.

---

[28] In Chap. 6 we see how one can remain (relatively) silent about the Problem of Ontology and yet say something substantial about the Problem of Handling.

[29] See Hughes' (1997, p. 330), Odenbaugh's (2015, pp. 281–82; 2018, Sect. 3), Teller's (2001a, p. 399), and Thomson-Jones' (2010).

[30] We note in passing that a similar issue arises in the literature on mental representation. As Cummins (1991, pp. 31–32) points out, if one were to think that mental states represented their targets in virtue of being similar to them, then one might be committed to the (absurd) conclusion that when you think about the cat on your lap your mental state is warm and furry.

A collection of properties is an abstract entity that is the sort of thing that can contain the properties specified by theoretical hypotheses as parts. The similarity relation between models and their targets shifts from the co-instantiation of properties, to the idea that targets instantiate (relevant) properties that are parts of the model. With respect to what it means for a model to be a bundle of properties Teller claims that the "[d]etails will vary with one's account of instantiation, of properties and other abstract objects, and of the way properties enter into models" (ibid.).

But, as Thompson-Jones (2010, pp. 294–95) notes, it is not obvious that this suggestion is an improvement on Giere's abstract objects. A bundle view incurs certain metaphysical commitments, chiefly the existence of properties and their abstractness; and a bundle view of objects, concrete or abstract, faces a number of serious problems (Armstrong 1989, Chap. 4). One might speculate that addressing these issues would push Teller either towards the kind of more robust account of abstract objects that he endeavoured to avoid, or towards a fictionalist understanding of models.

Fictionalists point out that a natural response to Teller's and Thomson-Jones' problem is to regard models as akin to *imaginary* or *fictional* systems of the sort presented in novels and films. It seems true to say that Sherlock is a smoker, despite the fact that Sherlock is an imaginary detective, and smoking is a physical property. At times, Giere seems sympathetic to this view. He says "it is widely assumed that a work of fiction is a creation of human imagination [...] the same is true of scientific models. So, ontologically, scientific models and works of fiction are on a par. They are both imaginary constructs" (2009, p. 249), and he observes that "novels are commonly regarded as works of imagination. That, ontologically, is how we should think of abstract scientific models. They are creations of scientists' imaginations. They have no ontological status beyond that" (2010, p. 278). However, these seem to be occasional slips and he recently positioned himself as an outspoken opponent of any approach to models that likens them to literary fiction. We discuss these approaches as well as Giere's criticisms of them in Sect. 6.8.

In sum, the similarity view is yet to be equipped with a satisfactory account of the ontology of models, and it has to be augmented with an account of their handling. Finally then, to round off the discussion it should also be noted that the similarity view has little to say about the Applicability of Mathematics Condition. Presumably this condition is tied up with the Problem of Ontology; in whatever way that models should be thought of ontologically speaking, they should be the sorts of things that can be specified by mathematical equations (since these are often part of the model-descriptions used to specify them), but they needn't be identified with mathematical structures per se (a position discussed in the next chapter). Keeping the Applicability of Mathematics Condition in mind then, places yet another constraint on the Problem of Ontology.

Chapter 4
The Structuralist View

The structuralist view of scientific representation originated in the so-called semantic view of theories, which came to prominence in the second half of the twentieth century.[1] The semantic view was originally proposed as an account of theory structure rather than an attempt to address questions of representation, but it is often taken to imply a view concerning scientific representation. The leading idea behind the position is that scientific theories are best thought of as collections of models. This invites the questions: what are these models, and how do they represent their target systems? Many defenders of the semantic view of theories take models to be *structures*, which represent their target systems in virtue of there being some kind of *mapping* (isomorphism, partial isomorphism, homomorphism, ...) between the two.[2] The idea that models are structures that represent by dint of a mapping between model and target is the core of the structuralist view of representation.

These structures can be thought of as mathematical objects: sets, or domains, with relations extensionally defined on them. We are familiar with many of these structures from our studies of mathematics. For example, $\langle \mathbb{N}, \leq \rangle$, the set of the natural numbers with the relation *less than or equal to*, is a structure in the sense relevant here. Readers who have studied first-order logic will also have come across these

---

[1] A discussion of the semantic view is beyond the scope of this book. Da Costa and French (2003), Suppes (2002), and van Fraassen (1980) provide classical statements of the view. A summary of the development of the view can be found in Suppe's (1989, pp. 3–37). Bailer-Jones (2009, Chap. 6), Byerly (1969), Chakravartty (2001), Klein (2013), and Portides (2005, 2010) provide critical discussions.

[2] Giere is sometimes also associated with the semantic view of theories despite not subscribing to either of these positions. We discussed his account of scientific representation in the previous chapter. Similarly, Suárez and Pero (2019) offer a variant of the semantic view, "the representational semantic conception", which also denies these claims whilst accepting that theories are, in some sense, collections of models (construed as representations along the lines of the account of representations we discuss in Sects. 5.2 and 5.3). See also Downes' (1992), Le Bihan's (2012), and Suppe's (1989) for relevant discussions.

structures when thinking about the semantics of the logic. In that context, structures provide the (codomain of) possible *interpretations* of formulae in the logic by assigning a domain to the universal quantifier, and by assigning subsets of the (n-fold) Cartesian product of that domain to the (n-place) predicate symbols of the logic.[3] Two such structures can be related to one another by functions. If those functions act so that they preserve the relations of the initial set (in the sense that their images are related in the appropriate way by the relations in the latter set), then those functions are structure-preserving. For example, if we think of an initial segment of the natural numbers, and a function which embeds that segment into the entire natural number structure, then we can see that the initial structure is preserved by the function (in this case an embedding). We will clarify the details of these notions later in the chapter, but for now it's worth bearing in mind that these structures are the sorts of homely mathematical objects that one encounters in studies of mathematics and logic.

This conception has two prima facie advantages. The first advantage is that it offers an account of epistemic representation that explains how surrogative reasoning works: the mappings between the model and the target allow scientists to convert truths found in the model into claims about the target system. The second advantage concerns the applicability of mathematics. There is time-honoured position in the philosophy of mathematics which sees mathematics as the study of structures; see, for instance, Resnik's (1997) and Shapiro's (2000). It is a natural move for the scientific structuralist to adopt this point of view, which then provides a neat explanation of how mathematics is used in scientific modelling.

We begin this chapter with a discussion of the issue of demarcation (Sect. 4.1). Whilst there has been little explicit discussion regarding how the structuralist approaches the two demarcation problems, related discussions concerning the structure of scientific theories and the structuralist position's close affinity to *mathematical* epistemic representations can be brought to bear in this context. We then move onto a more detailed discussion of what *structures* are, and thus the Problem of Carriers (Sect. 4.2). Next, we investigate how an appeal to structure-preserving mappings deals with epistemic representation (Sect. 4.3). We then turn to the Problem of Style and the Problem of Accuracy (Sect. 4.4). We end the chapter with a discussion of a puzzle that is particular to the structuralist conception: if scientific models are supposed to share "the structure" of their target systems, then where does this target end structure come from (Sect. 4.5)?

---

[3] In Sect. 4.2 we add a qualification to this.

4.1 The Demarcation Problems

Structuralism's stand on the two demarcation problems is by and large an open question. Unlike similarity, which has been widely discussed across different domains, the structures invoked by the structuralist view of scientific representation are tied closely to the formal framework of set theory, and the account has been discussed only sparingly outside the context of the mathematized sciences. It seems plausible, however, to assume that structuralists give a positive answer to the Taxonomic Representational Demarcation Problem and endorse at least the distinction between direct and indirect representation that we introduced in Sect. 3.1, and possibly also further distinctions. Sometimes scientists offer straightforward descriptions of the things they are interested in. This happens, for instance, when a geologist offers an account of the formation of the Alps by pointing out that they were created when the African Plate collided with the Eurasian Plate, which caused the rocks to be folded up to form a mountain range. It equally happens when air crash investigators present their report in which they describe in detail how a defective rudder made it impossible for the pilots to control the plane and how this led to the disaster. In both cases we are presented with direct description of the subject matter of interest. By contrast, when engaging in modelling, scientists first introduce a secondary object, the model, which they then take to represent the target. In the current context this secondary object is a mathematical structure, which, like the non-material models we discussed in Sect. 3.1, is usually specified by a description. Indeed, it is one of the core posits of the semantic view that structures themselves, and not the varying descriptions that we can give of these structures, are the units that represent a theory's target systems (see, for instance, van Fraassen's (1980)). In the last chapter the representation relation between model and target was assumed to be based on the model being similar to the target. In this chapter we explore the notion that representation is based on there being a suitable structure-preserving mapping between model and target. Thus, there is a clear difference between how the model–target relation is understood in the two approaches, but it leaves the distinction between direct and indirect representation intact.

The question now is whether the structuralist conception makes further taxonomic distinctions within the class of indirect representations. There are several options, ranging from humble to bold. A humble way of responding to the Taxonomic Representational Demarcation Problem introduces further distinctions. Within the class of indirect representations, the structuralist can, for instance, distinguish between mathematical models and other kinds of indirect representations, and choose to focus solely on mathematical models because these are most obviously characterised as structures. For example, the Newtonian model of the orbits discussed in the introduction can be characterised as the mathematical structure satisfying the differential equation $\ddot{\vec{x}} = -Gm_s \vec{x} / |\vec{x}|^3$, and therefore naturally seems to be within the purview of a structuralist account. This amounts to recognising mathematical models as a type of indirect representation that is distinguished from other types of indirect representation such as diagrams, photographs, and so on.

Focussing on mathematical models de facto involves demarcating even within the class of models themselves, namely by drawing a distinction between mathematical and non-mathematical models. In the latter group one would find material models like the Phillips–Newlyn machine or the San Francisco Bay model because material objects of this kind don't have a structure, at least not in any obvious way (we discuss the problem of what it means for physical object to have a structure later in Sect. 4.5). But the class of non-mathematical models need not only contain material models. Indeed, one might argue that non-mathematised sciences work with models that aren't structures. Godfrey-Smith (2006), for instance, argues that models in many parts of biology are imagined concrete objects. So the structuralist might not attempt to explain how models of that kind represent.[4]

A second, more ambitious, way of responding to the Taxonomic Representational Demarcation Problem attempts to apply the structuralist account to all models, specifically also to non-mathematical models. The taxonomy would then only distinguish between models and non-models, but not between different kinds of models. One worry facing this position is that this requires a rational reconstruction of scientific modelling, and as such it has some distance from the actual practice of science. Some philosophers have worried that this distance is too large, and that such a view is too far removed from the actual practice of science to be able to capture what matters to the practice of modelling,[5] because even though some models may be best thought of as structures, there are many where this seems to contradict how scientists actually talk about, and reason with, them.

Finally, a bold way of responding to the Taxonomic Representational Demarcation Problem is for the structuralist to adopt the same generalist attitude as proponents of the similarity approach and take their account to apply to indirect epistemic representations across the board. This is suggested by French (2003), who presents Budd's (1993) account of pictorial representation in detail and points out that it is based on the notion of a structural isomorphism between the structure of the surface of the painting and the structure of the relevant visual field. Therefore, representation is isomorphism of perceived structure (French 2003, pp. 1475–76).[6] In a similar vein, Bueno claims that the partial structures approach offers a framework in which different representations – among them "outputs of various instruments, micrographs, templates, diagrams, and a variety of other items" (2010, p. 94) – can be accommodated. This would suggest that at least some proponents of the structuralist account of representation take it to have a claim to cover representations across

---

[4] Although see French's (2014, Chap. 12) and Lloyd's (1994, 1984) for attempts to deploy structuralism in biological contexts.

[5] This is part of the thrust of many contributions to Morgan and Morrison's (1999); see also Cartwright's (1999a).

[6] This point is reaffirmed by Bueno and French (2011, pp. 864–65). Downes discusses French's use of Budd's theory in his (2009, pp. 423–25).

4.1 The Demarcation Problems

different domains.[7] This approach faces the same questions as the previous options. First, neither a visual field nor a painting is a structure, and the notion that they, in some sense, "share a structure" needs at least unpacking. Second, Budd's theory is only one among many theories of pictorial representation, and most alternatives do not invoke the preservation of structure. So there is question whether a generalist claim can be built on Budd's theory.

In sum, we think that it is reasonable to assume that the structuralist conception is associated with a direct/indirect distinction. Beyond this there are different ways of thinking about the structuralist conception of representation. We do not adjudicate between these options here. However, in the remainder of this chapter we will follow Bueno and French and discuss the structuralist conception in the bold version that aims to cover all indirect epistemic representations. Under this reading, the structuralist, like the similarity theorist, has to address the TR-Problem, which amounts to filling the blanks in two TR-schemes, namely "X is a direct scientific/epistemic representation of T iff ___" and "X is an indirect scientific/epistemic representation of T iff ___". As in the case of the similarity view, the analysis of direct representation can then be seen as the province of the philosophy of language and the structuralist can see her project as coming to grips with indirect representation. Those who prefer humble options will replace the above scheme with a scheme that mentions their own types, for instance they can address the problem formalised as "X is a mathematical model scientific/epistemic representation of T iff ___". Little, if anything, in our discussion in later sections will depend on this, and our arguments apply, mutatis mutandis, also to humble versions of the view.

To determine whether the indirect representation TR-scheme has to be completed for "scientific" or "epistemic", the structuralist has to address the Scientific Representational Demarcation Problem. As we have seen in the last chapter, proponents of the similarity view are mostly generalists and claim that all indirect representations are grounded in similarity. Proponents of the bold version do not demarcate between different indirect epistemic representations, and their lists of representations that are covered by the account include items that one would standardly classify as scientific (like models) and ones that one would not (like paintings). This suggests that they follow the similarity view's generalist attitude and don't demarcate between scientific and non-scientific representations. This means that structuralists, like proponents of the similarity view, end up filling the blank in "X is an indirect epistemic representation of T iff ___", which means that they address the Indirect Epistemic Representation Problem.

---

[7] Isaac (2019) discusses appeals to isomorphism in the analysis of perceptual experience, mental imagery, and visual artwork. These appeals underscore that at least in some theoretical contexts the isomorphism view has universal aspirations.

4.2 The Problem of Carriers

Almost anything from a concert hall to a kinship system can be referred to as a "structure". So the first task for a structuralist account of epistemic representation is to articulate what notion of structure it employs. A number of different notions of structure have been discussed in the literature (for a review see Thomson-Jones' (2011)), but by far the most common and widely used is the notion of structure one finds in set theory and mathematical logic. A structure S in that sense (sometimes "mathematical structure" or "set-theoretic structure") is a composite entity consisting of the following: a non-empty set U of objects called the domain (or universe) of the structure and a non-empty indexed set R of relations on U.[8] It is convenient to write these as $S = \langle U, R \rangle$, where "$\langle \, , \, \rangle$" denotes an ordered tuple. Sometimes operations are also included in the definition of a structure. While convenient in some applications, operations are redundant because operations reduce to relations.[9]

It is important to be clear on what we mean by "object" and "relation" in this context.[10] In defining the domain of a structure it is irrelevant what the objects are. All that matters from a structuralist point of view is that there are so and so many of them. Whether the object is a desk or a planet is immaterial. All we need are dummies or placeholders whose *only* property is "objecthood". Similarly, when defining relations one disregards completely what the relation is "in itself". Whether we talk about "being the mother of" or "standing to the left of" is of no concern in the context of a study of structures; all that matters is between which objects it holds. For this reason, a relation is specified purely extensionally: as a class of ordered n-tuples. The relation literally is nothing over and above this class. So a structure consists of dummy-objects between which purely extensionally defined relations hold.

Let us illustrate this with an example. Consider the structure with the domain $U = \{a, b, c\}$ and the following two relations: $r_1 = \{a\}$ and $r_2 = \{\langle a, b \rangle, \langle b, c \rangle, \langle a, c \rangle\}$. Hence R consists of r_1 and r_2, and the structure itself is $S = \langle U, R \rangle$. This is a structure with a three-object domain endowed with a monadic property and a transitive relation. Whether the objects are books or iron rods is of no relevance to the structure; they could be literally anything one can think of. Likewise r_1 could be any monadic property (being green, being waterproof, etc.) and r_2 could be any (irreflexive, anti-symmetric) transitive relation (larger than, hotter than, more expensive than, etc.).

It is worth pointing out that this use of "structure" differs from the use one sometimes finds in first-order logic, where linguistic elements are considered part of the

---

[8] With the exception of the caveat below regarding interpretation functions, this definition of structure is widely used in mathematics and logic. See, for instance, Hodges' (1997, p. 2), Machover's (1996, p. 149), and Rickart's (1995, p. 17). Following common practice, we're here focussing on first-order structures. For a discussion of higher-order structures see Hudetz's (2019). The arguments we make in this chapter equally apply to higher-order structures.

[9] See, for instance, Boolos and Jeffrey's (1989, pp. 98–99).

[10] For a clear account of the nature of structures we refer the reader to Russell's (1919/1993, p. 60).

4.2 The Problem of Carriers

carrier of the representation, or, for those who prefer to frame the problem more narrowly, the model. Specifically, over and above $S = \langle U, R \rangle$, in that context a structure is also taken to include a language L, sometimes called a *signature*, and an *interpretation function* assigning objects of S to elements of the signature.[11] A signature can be thought of as containing the universal quantifier ∀ and n-place predicate symbols (we're suppressing a discussion of constant and function symbols). An interpretation function assigns U to the universal quantifier, and each n-place predicate is assigned an n-place relation in R. In this way, we can talk about structures as *satisfying* or *making true* sentences in a first order logic. For example, consider a signature consisting of only the universal quantifier and a one-place predicate symbol P, a structure $S = \langle U, \{r\} \rangle$ such that $r = U$, and an interpretation function that assigns U to the universal quantifier and r to the predicate symbol P. Then, that structure, under that interpretation, makes the sentence "$\forall x P x$" true.

The distinction between thinking of structures as defined only in terms of a domain and relations, and thinking about structures as also including interpreting linguistic elements is subtle.[12] Proponents of the semantic view have traditionally insisted on avoiding yoking a scientific theory to a particular syntax or signature (cf. van Fraassen's (1989, p. 366)) and worked with a definition of structure in terms of a domain and relations. For our current purposes the difference between the two conceptions of structure is largely immaterial and so we stick with a version that takes a structure to be an ordered pair $S = \langle U, R \rangle$ as introduced above, and we disregard this alternative use of "structure" tied up with a signature.

The first basic posit of the structuralist theory of representation is that carriers are structures in the sense just discussed (the second is that they represent their targets by being suitably morphic to them; we discuss morphisms in the next section). Suppes, working with version of the view that focuses on models, articulated this stance clearly when he declared that "the meaning of the concept of model is the same in mathematics and the empirical sciences" (1969a, p. 12). Likewise, van Fraassen posits that a "scientific theory gives us a family of models to represent the phenomena", that "[t]hese models are mathematical entities, so all they have is structure" (1997, pp. 528–29) and that therefore "[s]cience is [...] interpreted as saying that the entities stand in relations which are transitive, reflexive, etc. but as giving no further clue as to what those relations are" (1997, p. 516). Redhead submits that "it is this abstract structure associated with physical reality that science aims, and to some extent succeeds, to uncover" (2001, p. 75). Finally, French and Ladyman affirm that "the specific material of the models is irrelevant; rather it is the structural representation [...] which is important" (1999, p. 109).[13]

---

[11] See, for instance, Enderton's (2001, pp. 80–81), Hodges' (1997, Chap. 1), and Machover's (1996, Chap. 8).

[12] See the relevant discussions in Glymour's (2013), Halvorson's (2012, 2013), and van Fraassen's (2014), as well as a useful overview in Lutz's (2017).

[13] Further explicit statements of this view are offered in Da Costa and French's (1990, p. 249), Suppes' (1969b, p. 24; 1970, Chap. 2 p. 6, 9, 13, 29) and van Fraassen's (1980, p. 43, 64; 1991, p. 483; 1995, p. 6; 1997, p. 516, 522).

As noted in the introduction to this chapter, these structuralist accounts have typically been proposed in the framework of the so-called semantic view of theories. There are differences between them, and formulations vary from author to author. However, as Da Costa and French (2000) point out, all these accounts share a commitment to analysing carriers of an indirect representation as structures. So we are presented with a clear answer to the Problem of Ontology: carriers are structures. The remaining issue is what structures themselves are, but philosophers of science can pass off the burden of explanation to philosophers of mathematics. This is what usually happens, and hence we don't pursue this matter further.[14]

The Problem of Handling is a little more pertinent, although it hasn't been discussed in much detail in the literature. As introduced earlier, the Problem of Handling includes the question of how scientists get epistemic access to the carriers of representation. Again, one might think that this question can be passed off to philosophers of mathematics and mathematicians: we learn about mathematical structures in science in the same way that we learn about mathematical structures in general. But as Nguyen (2017) notes, this can depend on the language used to present the structures, and at least in its canonical formulation, linguistic signatures aren't considered part of structures in the structuralist view. Thus, how the view accommodates at least some aspects of the Problem of Handling remains an open question for the structuralist.

An extension of the standard conception of structure is the so-called partial structures approach (see, for instance, Da Costa and French's (2003) and Bueno et al. (2002)). Above we defined an n-place relation by specifying which n-tuples belong to the relation. This naturally allows a sorting of all n-tuples into two classes: ones that belong to the relation and ones that don't. Hence, for any n-ary relation r we can sort the class of n-tuples into two disjoint sets, a set $r^{\in}$ of n-tuples that belong to a relation and set $r^{\neq}$ of the tuples that do not. The leading idea of partial structures is to introduce a third option: for some n-tuples it is indeterminate whether or not they belong to the relation. Rather than separating all n-tuples into two classes we now separate n-tuples into three classes: n-tuples that belong to the relation, n-tuples that do not belong to the relation, and ones for which it is indeterminate whether they belong to the relation or not. Let us denote the last class by "r^i". So while an "ordinary" relation is defined by $r^{\in}$ alone, a *partial relation* r is defined by the triple $(r^{\in}, r^{\neq}, r^i)$. These three classes are mutually exclusive (no n-tuple can be in more than one) and jointly exhaustive (every n-tuple must belong to one of the three classes). In the context of the partial structures approach, "ordinary" relations are also referred to as *total relations*.

If the set R of a structure contains at least one partial relation, then the structure is a *partial structure*. Formally, a partial structure S_p is tuple $\langle U, R_p \rangle$ where U is a domain of objects and R_p is an indexed set containing at least one partial relation on

---

[14] The ontological status of structures is hotly debated in the philosophy of logic and mathematics: are they Platonic entities, equivalence classes, modal constructs, or yet something else? Dummett (1991, 295ff.), Hellman (1989), Redhead (2001), Resnik (1997), and Shapiro (2000) offer different views on this issue.

4.2 The Problem of Carriers

U. An "ordinary" structure, that is one with no partial relations, is called a *total structure*.

Partial structures make room for a process of scientific investigation where one begins not knowing whether an n-tuple falls under the relation and then learns whether or not it does.[15] In many cases we know that a relation applies to certain objects; we also know that it does not apply to other objects; but there are a number of objects of which we simply do not know whether or not the relation applies to them. So the three classes that define a partial relation can be given an epistemic interpretation. Under this interpretation, partial relations offer a representation of the incompleteness of our knowledge and capture that openness of scientific theories to new developments. The relations in r^i suggest lines of inquiry aiming to find out whether certain n-tuples can be reclassified as belonging to either $r^\in$ or $r^{\not\in}$.

While the extension of structures to partial structures may have advantages when it comes to thinking about scientific progress or research strategies, it has no ontological implications because partial structures are ontologically on par with ordinary (or total) structures. It turns out, however, that as least some proponents of the partial structures approach are more guarded as regards the ontology of carriers (although we note that the connection is entirely contingent: the same precautionary attitude could be adopted towards total rather than partial structures). Bueno and French emphasise that "advocates of the semantic account need not be committed to the ontological claim that models *are* structures" (2011, p. 890 original emphasis). This claim is motivated by the idea that the task for philosophers of science is to represent scientific theories and models, rather than to reason about them directly. French (2010) makes it explicit that according to his version of the semantic view of theories, a scientific theory is *represented* as a class of models, but should not be identified with that class (this point reaffirmed in his (2017) and used to address the discussions highlighted in footnote 12). Moreover, a class of models is just one way of representing a theory; we can also use an intrinsic characterisation and represent the same theory as a set of sentences in order to account for how they can be objects of our epistemic attitudes (French and Saatsi 2006).[16]

French therefore adopts a quietist position with respect to what a theory or a model *is*, declining to answer the question (2010, 2017; cf. French and Vickers 2011). There are thus two important notions of representation at play: a primary representation of targets by carriers, which is the job of scientists, and a secondary representation of theories and carriers by structures, which is the job of philosophers of science. The question for this approach then becomes whether or not the structuralist representation of carriers and epistemic representation – as partial structures and morphisms that hold between them – is an accurate or useful one.

---

[15] The partial relations bear an intimate relation to Mary Hesse's classification of analogies as positive, negative, and neutral (Hesse 1963, Chap. 2).

[16] This would make the semantic view of theories compatible with views that identify a theory with linguistic elements, in particular earlier versions of structural realism such as Worrall's (1989), which themselves, at least in part, motivated the position.

And the concerns raised in the later sections of this chapter remain when translated into this context as well.

There is an additional question regarding the correct formal framework for thinking about carriers in the structuralist position. Landry (2007) argues that in certain contexts group theory, rather than set theory, should be used when talking about structures and morphisms between them (see also Roberts's (2010)). However, the details of this debate do not matter too much for our current project: Landry is not attempting to reframe the representational relationship between carriers and their targets[17] and the question of how individual structures represent remains much the same irrespective of which framework is used (see French's (2014, Chaps. 5, 6) for a relevant discussion).

4.3 The Indirect Epistemic Representation Problem

The most basic structuralist conception of scientific representation asserts that scientific models, understood as structures, represent their target systems in virtue of being isomorphic to them. Two structures $S_a = \langle U_a, R_a \rangle$ and $S_b = \langle U_b, R_b \rangle$ are isomorphic iff there is a mapping $f: U_a \to U_b$ such that (i) f is a bijection (i.e. a one-to-one correspondence), and (ii) f preserves the system of relations in the following sense: the members $a_1, \ldots, a_n$ of U_a satisfy the relation r_a of R_a iff the corresponding members $b_1 = f(a_1), \ldots, b_n = f(a_n)$ of U_b satisfy the relation r_b of R_b, where r_b is the relation corresponding to r_a. U_a is called the *domain* of the mapping and U_b the *codomain*.[18] If $b = f(a)$ then b is called the *image* of a, and a is the *pre-image* of b. What it means for r_b to "correspond" to r_a is a subtle question. Recall that we defined the second element of a structure as an *indexed* set of relations defined over the domain. This means that the members of the set have an index and so we can speak, for instance, of the 5th relation in the set. Appealing to indices one can then say that corresponding relations are the ones that have the same index. In order for two structures to be isomorphic their relations have to be indexed so that condition (ii) is satisfied (notice that this has the consequence that permuting the order of the relations will in general yield a non-isomorphic structure). Another way of cashing out what it means for the relations to correspond is that the interpretation function associated with the structures assigned them to the same element of the language L; the relations both "interpret" the same syntactic sign.[19]

We are now in a position to discuss how the structuralist view deals with epistemic representation. As we have seen in Sect. 4.1, the structuralist view has to

---

[17] See Brading and Landry's (2006) for her scepticism regarding how structuralism deals with this question.

[18] Sometimes the codomain is also referred to as the *target domain*. We avoid this terminology in the current context to guard against conflating target domains (which are mathematical objects) and target systems (which are the parts of the world that a model represents).

[19] For more on this see Halvorson's (2012), Glymour's (2013), and Lutz's (2017).

4.3 The Indirect Epistemic Representation Problem

address the Indirect Epistemic Representation Problem. Assume now that the target system T exhibits the structure $S_T = \langle U_T, R_T \rangle$ and the carrier X is the structure $S_X = \langle U_X, R_X \rangle$. Then X represents T iff X is isomorphic to the structure exhibited by T:

Structuralism 1: a carrier X is an indirect epistemic representation of its target T iff S_X is isomorphic to S_T.

This view is articulated explicitly by Ubbink, who posits that "a model represents an object or matter of fact in virtue of this structure; so an object is a model [...] of matters of fact if, and only if, their structures are isomorphic" (1960, p. 302). Views similar to Ubbink's seem to be operable in many versions of the semantic view. In fairness to proponents of the semantic view it ought to be noted, though, that for a long time representation was not the focus of attention in the view and the attribution of (something like) Structuralism 1 to the semantic view is an extrapolation. Representation became a much-debated topic in the first decade of the twenty-first century, and many proponents of the semantic view then either moved away from Structuralism 1, or pointed out that they never held such a view. We turn to more advanced positions shortly, but to understand what motivates such positions it is helpful to understand why and how Structuralism 1 fails.

An immediate question concerns the introduction of a target end structure S_T. At least prima facie, target systems aren't structures; they are physical objects like planets, molecules, bacteria, tectonic plates, and populations of organisms. The relation between structures and physical targets is indeed a serious question and we will return to it in Sect. 4.5.[20] In this subsection we grant the structuralist the assumption that target systems are (or at least have) structures.

The first and most obvious problem is the same as with the similarity view: isomorphism is symmetrical, reflexive, and transitive, but epistemic representation isn't. This problem could be addressed by replacing isomorphism with an alternative mapping. Bartels (2006), Lloyd (1984), and Mundy (1986) suggest homomorphism; van Fraassen (1980, 1997, 2008) and Redhead (2001) favour isomorphic embeddings; advocates of the partial structures approach prefer partial isomophisms (Bueno 1997, 1999; Da Costa and French 1990, 2003; French 2000; French and Ladyman 1999); and Swoyer (1991) introduces what he calls Δ/Ψ–morphisms. We refer to these collectively as "morphisms" and discuss them in more detail in Sect. 4.4. This solves some, but not all problems. While many of these mappings are asymmetrical, they are still reflexive, and at least some of them are also transitive. But even if these formal issues could be resolved in one way or another, a view based on structural mappings would still face other serious problems. For ease of presentation we discuss these problems in the context of the isomorphism view; mutatis mutandis other formal mappings suffer from the same difficulties.

Like similarity, isomorphism is too inclusive: not all things that are isomorphic represent each other. In the case of similarity this was brought home by Putnam's

---

[20] An early recognition that the relation between targets and structures is not straightforward can be found in Byerly, who emphasizes that structures are abstracted from objects (1969, p. 135, 138).

thought experiment with the ant crawling on the beach; in the case of isomorphism a look at the history of science will do the job. Many mathematical structures have been discovered and discussed long before they have been used in science. Non-Euclidean geometries were studied by mathematicians long before Einstein used them in the context of spacetime theories, and Hilbert spaces were studied by mathematicians prior to their use in quantum theory. If representation was nothing over and above isomorphism, then we would have to conclude that Riemann discovered general relativity or that Hilbert invented quantum mechanics, which is obviously wrong. Isomorphism on its own does not establish representation (Frigg 2002, p. 10).

Isomorphism is more restrictive than similarity: not everything is isomorphic to everything else. But isomorphism is still too easy to come by to correctly identify the extension of a representation (i.e. the class of systems it represents), which gives rise to a version of the mistargeting problem. The root of the difficulties is that the same structures can be instantiated in different target systems. The $1/r^2$ law of Newtonian gravity is also the mathematical skeleton of Coulomb's law of electrostatic attraction and the weakening of sound or light as a function of the distance to the source.[21] The mathematical structure of the pendulum is also the structure of an electric circuit with condenser and solenoid.[22] Linear equations are ubiquitous in physics, economics, and psychology. Certain geometrical structures are instantiated by many different systems; just think about how many spherical things we find in the world. This shows that the same structure can be exhibited by more than one target system. Borrowing a term from the philosophy of mind, one can say that structures are *multiply realisable.* If representation is explicated solely in terms of isomorphism, then we have to conclude that, say, a model of a pendulum also represents an electric circuit. This seems wrong. Of course, the model *could* also be used to represent an electric circuit. But the fact that it could be used this way does not entail that it does represent the circuit. But the claim that it does is forced upon us by the idea that representation is established by isomorphism alone. Hence isomorphism is too inclusive to correctly identify a representation's extension.

One might try to dismiss this point as an artefact of a misidentification of the target van Fraassen (1980, p. 66) mentions a similar problem under the heading of "unintended realisations" and then expresses confidence that it will "disappear when we look at larger observable parts of the world". Even if there are multiply realisable structures to begin with, they vanish as science progresses and considers more complex systems because these systems are unlikely to have the same structure. Once we focus on a sufficiently large part of the world, so the thought goes, no two relevantly distinct phenomena will have the same structure. There is a problem with this counter, however. Why would we have to appeal to *future* science to explain how models work *today*? It is a matter of fact that we currently have models that represent electric circuits and sound waves, and we do not have to await future

---

[21] Nguyen (2017) argues that examples such as these put pressure on idea that theoretical equivalence can be explicated in formal terms without reference to how models are interpreted by model users.

[22] For a detailed discussion of this case see Kroes' (1989).

science providing us with more detailed accounts of a phenomenon to make our models represent what they actually already represent.

As we have seen in the last section, a misrepresentation is one that portrays its target as having features that the target doesn't have. In the case of an isomorphism account of representation this presumably means that the model portrays the target as having structural properties that it doesn't have. However, isomorphism demands identity of structure: the structural properties of the model and the target must correspond to one another exactly. A misrepresentation won't be isomorphic to the target. By the lights of Structuralism 1 it is therefore is not a representation at all. Like simple similarity accounts, Structuralism 1 conflates misrepresentation with non-representation.

Muller (2011, p. 112) suggests that this problem can be overcome in a two-stage process: one first identifies a sub-model of the model, which in fact is isomorphic to at least a part of the target. This "reduced" isomorphism establishes representation. One then constructs a "tailor-made morphism on a case by case basis" (ibid.) to account for accurate representation. Muller is explicit that this suggestion presupposes that there is "at least one resemblance" (ibid.) between model and target because "otherwise one would never be called a representation of the other" (ibid.). While this may work in some cases, it is not a general solution. It is not clear whether all misrepresentations have isomorphic sub-models. Models that are gross distortions of their targets (such as the liquid drop model of the nucleus or the logistic model of a population) may well not have such sub-models. More generally, as Muller admits, his solution "precludes total misrepresentation" (ibid.). So in effect it just limits the view that representation coincides with correct representation to a sub-model. However, this is too restrictive a view of representation. Total misrepresentations may be useless, but they are representations nevertheless.

Another response refers to the partial structures approach and emphasises that partial structures are in fact constructed to accommodate a mismatch between model and target and are therefore not open to this objection (Bueno and French 2011, p. 888). It is true that the partial structures framework enjoys a degree of flexibility that the standard view does not have.[23] However, we doubt that this flexibility stretches far enough. While the partial structures approach deals successfully with incomplete representations, it does not seem to deal well with distortive representations (we come back to this point in the next section). So the partial structures approach, while enjoying some advantages over the standard approach, is nevertheless not yet home and dry.

Like the similarity account, Structuralism 1 has a problem with non-existent targets because a model cannot be isomorphic to something that doesn't exist. If there is no ether, a model can't be isomorphic to it. Hence carriers without targets cannot represent what they seem to represent.

---

[23] Vickers (2009) discusses the question whether this flexibility goes as far as being able to accommodate inconsistent representations.

Most of these problems can be resolved by making moves along the lines of the ones that lead to Similarity 5: introduce agents and hypothetical reasoning into the account of representation. Going through the motions one finds:

Structuralism 2: a carrier X is an indirect epistemic representation of a target system T iff there is an agent A who uses X to represent a target system T by specifying a theoretical hypothesis H proposing an isomorphism between S_X and S_T (where S_X and S_T correspond to the structures of X and T, respectively).

Something like this was suggested by Adams (1959) who appeals to the idea that physical systems are the *intended* models of a theory in order to differentiate them from purely mathematical models of a theory. This suggestion is also in line with van Fraassen's recent pronouncements on representation. He offers the following as the "Hauptstatz" of a theory of representation: *"There is no representation except in the sense that some things are used, made, or taken, to represent things as thus and so"* (2008, p. 23, original emphasis). Likewise, Bueno submits that "representation is an *intentional* act relating two objects" (2010, p. 94, original emphasis), and Bueno and French point out that using one thing to represent another thing is not only a function of (partial) isomorphism but also depends on "pragmatic" factors "having to do with the use to which we put the relevant models" (2011, p. 885). This, of course, gives up on the idea of an account that reduces representation to intrinsic features of models and their targets. At least one extra element, the model user, also features in whatever relation is supposed to constitute the representational relationship between X and T.[24] In a world with no agents, there would be no epistemic representations.

This seems to be the right move. Like Similarity 5, Structuralism 2 accounts for the directionality of representation and has no problem with misrepresentation. But, again as in the case of Similarity 5, this is a Pyrrhic victory as the role of isomorphism has shifted. The crucial ingredient is now the agent's intention, and isomorphism has in fact become either a representational style or normative criterion for accurate representation. Let us now assess how well isomorphism fares as a response to these problems.

4.4 Accuracy and Style

The Problem of Style is to identify representational styles and to characterise them. Isomorphism offers an obvious response to this challenge: one can represent a system by coming up with a model that is structurally isomorphic to it. We call this the isomorphism-style. This style also offers a clear-cut condition of accuracy: the

---

[24] Although French and Bueno (2011, pp. 886–87) argue against this, claiming that even if scientific representation is context dependant – in the sense that it depends on the intentions of model users – we shouldn't write this into the answer to the Indirect Epistemic Representation Problem; cf. French's (2003, pp. 1473–74).

4.4 Accuracy and Style

representation is accurate if the hypothesised isomorphism holds; it is inaccurate if it doesn't.

One issue with this suggestion is that isomorphism is demanding: two structures are isomorphic iff they agree with respect to all of their structural features. Any mismatch between model and target will destroy an isomorphism. As we have previously seen, many models deviate in one way or another from reality and hence there are many cases of epistemic representation that don't match the structure of the target system in all respects. For this reason, defenders of the structuralist view have appealed to weaker morphisms too (see the references given earlier). To pave the ground for a discussion of such morphisms it is instructive to first recall the definition of isomorphism (which was stated in Sect. 4.3):

Two structures $S_a = \langle U_a, R_a \rangle$ and $S_b = \langle U_b, R_b \rangle$ (where the elements of both R_a and R_b share the same indices) are isomorphic iff there is a mapping $f: U_a \rightarrow U_b$ such that (i) f is a bijection (i.e. a one-to-one correspondence), and (ii) f preserves the system of relations in the following sense: the members $a_1, ..., a_n$ of U_a satisfy the relation r_a of R_a iff the corresponding members $b_1 = f(a_1), ..., b_n = f(a_n)$ of U_b satisfy the relation r_b of R_b, where r_b is the relation corresponding to r_a in the sense of sharing the same index.

The notion of an isomorphism can be weakened in different ways.[25] A mapping f is injective iff it never maps distinct elements of the domain to the same element of its codomain: $f(x) \neq f(y)$ for all $x \neq y$. A mapping f is surjective iff each element of the codomain is mapped to by at least one element of the domain: for every z in the codomain there exists an x in the domain so that $z = f(x)$. Colloquially speaking, a surjective mapping "hits" the entire codomain. A mapping is bijective iff it is injective and surjective. One can then replace the requirement that f be bijective with the weaker requirement that it only be injective. Dropping the requirement that the mapping be surjective therefore makes room for there being elements in U_b that are not in the image of the mapping. If this holds, we say that f is an *embedding* of $S_a = \langle U_a, R_a \rangle$ into $S_b = \langle U_b, R_b \rangle$. Clearly if an embedding is also surjective then it is an isomorphism. If a model is based on an embedding, one can say that represents in the "embedding style". It is accurate if the conditions for an embedding hold; it is inaccurate if they don't. Embeddings are particularly useful to capture cases in which the carrier has "more" structure than the target.[26] Sometimes this additional structure is referred to as "surplus structure" (Redhead 2001, pp. 79–83).[27]

---

[25] We note here that some of the following mappings have analogues in the context of the partial structures approach. See Bueno's (1997, 1999), Da Costa and French's (1990, 2003), French's (2000) and French and Ladyman's (1999) for details.

[26] See Redhead's (2001) and van Fraassen's (1980, 1997, 2008) for discussions of isomorphic embeddings in this context.

[27] However, whilst embeddings might be one way of characterising surplus structure, more recent discussions prefer to compare, via the use of category theory, the respective symmetries of the structures involved to explicate this (Nguyen et al. 2020; Weatherall 2016b), and it's unclear whether this captures the same notion of surplus as a non-surjective embedding.

This can be weakened further by also dropping the requirement that f be injective and at the same time replacing the biconditional in condition (ii) of the definition of isomorphism with a conditional. This defines a *homomorphism*.[28] Two structures $S_a = \langle U_a, R_a \rangle$ and $S_b = \langle U_b, R_b \rangle$, where R_a and R_b are indexed in the same way, are homomorphic iff there is a function $f: U_a \to U_b$ such that for all r_a of R_a: if $a_1, \ldots, a_n$ of U_a satisfy r_a then the corresponding members $b_1 = f(a_1), \ldots, b_n = f(a_n)$ of U_b satisfy the relation r_b of R_b, where r_b is the relation corresponding to r_a in the sense of sharing the same indices. Allowing for the mapping to be neither injective nor surjective allows there to be elements in U_b that are not in the image of f and it allows for there to be cases where distinct elements of the domain are mapped to the same element of the codomain. Replacing the biconditional with a conditional makes room for there to be cases where a relation holds between certain elements in U_b but the corresponding relation does not hold between the pre-images of these elements in U_a. The implications for style and accuracy are entirely parallel to the cases we have seen previously. If a model is based on a homomorphism, one can say that it represents in the "homomorphism style". It is accurate if the conditions for a homomorphism hold; it is inaccurate if they don't.

Each of the above morphisms relied on the idea that relations over the structures shared the same index set. This condition too can be weakened. For some structure $S_a = \langle U_a, R_a \rangle$, we could consider a *reduct structure* $S_a^- = \langle U_a, R_a^- \rangle$, where the index set of R_a^- is a subset of the index set of R_a, and the two agree for all elements in the former. This would correspond to "forgetting" some of the relations in R_a. We could then talk about morphisms between reducts of some structure $S_a = \langle U_a, R_a \rangle$ and a structure $S_b = \langle U_b, R_b \rangle$.[29] This would allow us to have relations in the target system which don't correspond to anything in the carrier of the representation (i.e. which are not represented at all).

In the context of discussing epistemic representation, Swoyer (1991) suggests a novel, and even weaker, kind of mapping. He suggests that we can take two subsets, Δ and Ψ, of R_a, at least one of which is required to be non-empty, and require that the function f from U_a to U_b "preserve" the relations in Δ and "counter-preserve" the relations in Ψ. Formally, two structures $S_a = \langle U_a, R_a \rangle$ and $S_b = \langle U_b, R_b \rangle$, whose relations share the same indices, are Δ/Ψ–morphic, for chosen $\Delta, \Psi \subseteq R_a$ iff there exists a function f from U_a to U_b such that (i) for all $r_a \in \Delta$: if $a_1, \ldots, a_n$ of U_a satisfy r_a then the corresponding members $b_1 = f(a_1), \ldots, b_n = f(a_n)$ of U_b satisfy the relation r_b of R_b (where r_b shares the same index as r_a); and (ii) for all $r'_a \in \Psi$: if

---

[28] See Bartels' (2006), Lloyd's (1984), and Mundy's (1986) for discussions of homomophisms. Pero and Suárez (2016) provide a critical discussion of homomorphisms in the context of scientific representation.

[29] Note that if one were to take such an approach one would have to be very careful about defining how the relations in each structure were supposed correspond to one another. Reduct structures are usually discussed in the context where the structures interpret a signature, L and a sub-signature L^-, $L^- \subseteq L$, in which case things are straightforward. Each of the above notions are discussed extensively in the context of model theory. See, for example, Bell and Machover's (1977) and Hodges' (1997, Chap. 1), as well as Enderton's (2001) – although note that Enderton defines a homomorphism with a biconditional rather than a conditional.

4.4 Accuracy and Style

$b_1 = f(a_1), \ldots, b_n = f(a_n)$ of U_b satisfy the relation r'_b of R_b (again where r'_b and r'_a share the same index), then $a_1, \ldots, a_n$ satisfy r'_a.[30] Thus a Δ/Ψ morphism allows that only some of the structure defined over U_a be preserved (i.e. the relations in Δ), and that only some of the structure defined over U_b correspond to a structure defined over U_a (i.e. the relations corresponding to the image of the relations in Ψ).

These are but the most common structure-preserving mappings; other maps exist and yet others can be defined if the need arises. The structuralist view of epistemic representation thus has a wealth of different structure-preserving mappings that it can utilise in different contexts. The question is whether *all* epistemic representations (or at least all mathematical epistemic representations) are based on structure-preserving mappings. The idea that all epistemic representations can be explicated in terms of preservation of structure faces an immediate challenge, namely that the claim seems to conflict with scientific practice. As we have seen previously, partial structures and various weaker morphisms are well equipped to deal with incomplete representations. However, not all inaccuracies are due to something being left out. Some representations distort, deform, and twist properties of the target in ways that are difficult to capture in terms of structure preservation. Moreover, these distortions can play a crucial role in the carriers being able to accurately represent what they do. The Ising model contains an infinite number of particles, and it has to if it is to undergo a phrase transition. In order for the Newtonian model of a celestial orbit to yield results we idealised the model by taking the sun and the orbiting planet to be perfect spheres with a homogeneous mass distribution. And so on. Certain idealising assumptions introduce explicit distortions, and it is at best unclear how structure-preserving morphism can account for these kinds of idealisations.[31]

There are, then, two ways for a proponent of the structuralist view of epistemic representation to proceed. First, they can endeavour to defuse first impressions and give an explicit account of how structure-preserving morphisms can accommodate carriers that explicitly distort their target systems, where the distortion is such that it plays a positive epistemic role despite the carrier/target mismatch. Second, they may accept that there are other styles of epistemic representation beyond structure preservation and concede that structure-preserving morphisms must exists alongside other kinds of representations. We take the latter to be the more plausible option, but ultimately only time will tell whether, and if so how, structure-preserving morphisms can be utilised to capture distorting idealisations.

The previous discussion focused on how different morphisms preserve, in an intuitive sense, different amounts of structure. Once we see how mathematical models are used in scientific practice itself, where the explicit set-theoretic definition may not be given, there is an additional question concerning what the relevant

---

[30] Swoyer is working with what he calls "intensional relational systems" which aren't the same as the set-theoretic structures as defined here. But the difference is not relevant for our current purposes.

[31] See Pincock's (2005) for a discussion of this in the partial structures framework and Pero and Suárez's (2016) for a discussion which focuses on homomorphism. For a discussion of these idealisations in terms of limits see Sect. 9.3.

structure being used representationally is in the first place. For example, if someone is using $\mathbb{R}^4$ as an epistemic representation, are they using the metric on $\mathbb{R}^4$, or just the topology? Often this is clear enough in a given context, but more complicated situations can arise where there can be substantial disagreement concerning what the relevant structure of the mathematical model is.

There has been a surge of recent work in the philosophy of physics concerning how we might use the tools of category theory to represent physical theories and thereby address questions concerning theoretical equivalence and related inter-theory comparisons pertaining to notions such as surplus structure that can be brought to bear in this context. Restricting ourselves to the philosophical literature, the approach has its origins in Halvorson's (2012) and Weatherall's (2016a).[32] An extended discussion of these issues goes beyond the scope of this book. But, whilst these debates have remained somewhat disconnected from the literature on scientific representation (with the exception of Fletcher's (2020) and Nguyen's (2017)), they do raise issues that are worth briefly mentioning here.

The crucial idea behind analysing scientific theories in terms of categories is that the representational content of a theory, even according to (a modified version of) the semantic view of theories, should be sensitive to more than just the features of the models, still conceived of as mathematical structures, of the theory. The representational content should also be sensitive to the *relationships* between the models. This is because, at the very least, by focusing on the relationships between the models, we are able to isolate the relevant features of the structures, for example the metric or the topology.

Here category theory provides an appropriate framework to work in, since it allows us to explicitly formalise these relationships. A *category* consists of a collection of *objects*, and for each pair of objects x and y in that collection a collection hom(x, y) of *morphisms* or *arrows* from x to y. These morphisms are required to compose in the sense that for any morphism f in hom(x, y) and any morphism g in hom(y, z) (i.e. any pair of morphisms such that one is to an object y and the other is from an object y), there is a morphism gf ('g following f') in hom(x, z), and composition is associative in the sense that for any f in hom(x, y), g in hom(y, z), and h in hom(z, w): $h(gf) = (hg)f$. Finally, each object x in a category has an associated *identity arrow* 1_x such that for any morphism f in hom(x, y), $f1_x = f = 1_y f$.[33] As examples consider: the category SET whose objects are sets, and whose morphisms are functions; the category TOP whose objects are topological spaces, and whose morphisms are continuous maps; and the category VECT$_k$ whose objects are vector spaces over a field k, and whose morphisms are linear maps.

In the context of thinking about the structure of scientific theories, the idea is that we should think of them as categories whose objects are the theory's models,

---

[32] For a survey of the recent literature on the topic see Weatherall's (2019a, b).

[33] For introductions to category theory see, for example Leinster's (2014) and Mac Lane's (1998). For a philosophical discussion of category theory see the contributions to Landry's (2017).

4.4 Accuracy and Style

and whose morphisms encode the relationships between the models. Introducing the machinery of category theory provides us with the tools to think about various different kinds of inter-model relationships, but most relevant for our current purposes are those that formalise the "standards of sameness" for the models in question. This can be done by focusing on *invertible* morphisms of the category in question. In a category where f is a morphism in $\text{hom}(x, y)$, f is invertible if there is a morphism g in $\text{hom}(y, x)$ such that $gf = 1_x$ and $fg = 1_y$. These invertible morphisms are isomorphisms in the category in question (in this sense they correspond to the notion of isomorphism discussed above, and they make explicit what the relevant structure of the models in question is, relative to the category in question): in SET they are the bijective functions, in TOP they are the homeomorphisms, and so on. The idea is that for some mathematical theories, we have isomorphisms between the models that *constrain* the representational content of the models.[34] In particular, how we go about choosing the appropriate category with which to represent our physical theories, and how we specify the representationally relevant features of the mathematical objects in question, requires choosing a "standard of sameness" as provided by the relevant notion of isomorphism between each theory's models. Once two structures are judged to be isomorphic according to the category used, then they share the same representational content, since they should be interpreted in the same way.[35]

Now, whilst any account of scientific representation should be sensitive to these sorts of questions, the underlying issue of how individual models, again still conceived of as mathematical structures, represent remains unchanged. Only once we have the tools to answer that question can we settle the matter as to whether two models are interpreted in the same way. As such our strategy here will be to continue focusing on how individual models represent, leaving question of how inter-model relationships contribute for future research.

---

[34] Examples of such relations include isometries between diffeomorphic Lorentzian manifolds in general relativity (Weatherall 2018) and gauge transformations between one-forms on a differentiable manifold in U(1) electromagnetism (Nguyen et al. 2020; Weatherall 2016a, b). For more on the role of symmetries in interpreting physical theories see, for example, Caulton's (2015) and Dewar's (2019).

[35] It bears noting here that the category-theoretic framework could also be used to investigate additional questions that pertain to the representational capacity of a theory and its models. For example, we could also investigate how the *embedding* of one structure in another could be used to represent sub-system relationships between targets. One might also take the arrows of a category to play a more positive representational role, beyond merely constraining how the models are interpreted. For example, Nguyen et al. (2020) discuss the essential role that gauge transformations play in ensuring that field configurations on a manifold are represented as local in nature, in a way that goes beyond merely ensuring that gauge related one-forms are interpreted in the same way.

4.5 The Structure of Target Systems

Target systems are objects: atoms, planets, populations of rabbits, economic agents, markets, etc. Isomorphism is a relation that holds between two structures, and the same is of course true of any other morphism. Claiming that a set-theoretic structure is morphic to a piece of the physical world is prima facie a category mistake. By definition, morphisms only hold between two structures. If we are to make sense of the claim that a carrier is morphic to its target we have to assume that the target exhibits a certain structure $S_T = \langle U_T, R_T \rangle$. But what does it mean for a target system – a part of the physical world – to possess a structure, and where in the target system is such a structure located?

The two prominent suggestions in the literature are that data models are the target end structures represented by carriers, and that structures are, in some sense, instantiated in target systems.[36] The latter option comes in three versions. The first version is that a structure is ascribed to a system; the second version is that systems instantiate structural universals; and the third version claims that target systems simply are structures. Let us now have a closer look at all these suggestions.

What are data models? Data are what we gather in experiments. When observing the motion of the moon, for instance, we choose a coordinate system and observe the position of the moon in this coordinate system at consecutive instants of time. We then write down these observations. The data thus gathered are called the *raw* data. The raw data then undergo a process of cleansing, rectification, and regimentation: we throw away data points that are obviously faulty, take into consideration what measurement errors there are, and usually idealise the data, for instance by replacing discrete data points by a continuous function and subject them to statistical processes. Often, although not always, the result is a curve through the data points that satisfies certain theoretical desiderata.[37] These resulting data models can be treated as set-theoretic structures. In many cases the data points are numeric and the data model is a relation over $\mathbb{R}^n$ (for some n), or subsets thereof, and hence it is a structure in the requisite sense.

Suppes (1969b) was the first to suggested that data models are the targets of scientific models: models don't represent parts of the world; they represent data structures. This approach has then been adopted by van Fraassen, when he declares that "[t]he whole point of having theoretical models is that they should fit the phenomena, that is, fit the models of data" (1981, p. 667). He has defended this position numerous times over the years (1980, p. 64; 1985, p. 271; 1989, p. 229; 1997, p. 524; 2002, p. 164) including in his most recent book on representation (2008,

---

[36] Notice that this discussion of instantiation also applies to the question of how non-mathematical carriers, including material models, exhibit a "structure". Thus, the structuralist who intends her account to apply to such carriers will have to answer the question twice over: what is "the structure" of the target system and what is "the structure" of the carrier?

[37] See Harris's (2003) and van Fraassen's (2008, pp. 166–68) for further elaboration on this process. Leonelli (2016, 2019) offers detailed discussion of the acquisition and use of data in biology.

4.5 The Structure of Target Systems

p. 246, 252). On this view, then, carriers don't represent planets, atoms, or populations; they represent data that are gathered when performing measurements on planets, atoms or populations.

This revisionary point of view has met with stiff resistance. Muller articulates the unease about this position as follows: "the best one could say is that a data structure D seems to act as *simulacrum* of the concrete actual being B [...] But this is not good enough. We don't want simulacra. We want the real thing. Come on." (2011, p. 98). Muller's point is that science aims (or at least should aim) to represent real systems in the world and not data structures. van Fraassen calls this the "loss of reality objection" (2008, p. 258) and accepts that the structuralist must ensure that carriers represent target systems, rather than finishing the story at the level of data. In his (2008) van Fraassen addresses this issue in detail and offers a solution. We discuss his solution below, but before doing so we want to articulate the objection in more detail. To this end we briefly revisit the discussion about phenomena and data which took place in the 1980s and 1990s.

Bogen and Woodward (1988), Woodward (1989), and more recently (and in a somewhat different guise) Teller (2001b), articulated a distinction between phenomena and data and argued that carriers represent phenomena, not data. The difference is best introduced with an example: the discovery of weak neutral currents (Bogen and Woodward 1988, pp. 315–18). The carrier in question is a model of particles: neutrinos, nucleons, and the Z^0 particle, along with the interactions between them.[38] Nothing of that, however, shows in the relevant data. CERN in Geneva produced 290,000 bubble chamber photographs of which roughly 100 were considered to provide evidence for the existence of neutral currents. The notable point in this story is that there is no part of the model (provided by quantum field theory) that could be claimed to be isomorphic to these photographs. Weak neutral currents are the phenomenon under investigation; the photographs taken at CERN are the raw data, and any summary one might construct of the content of these photographs would be a data model. But there is nothing in the model that could be meaningfully claimed to represent the data model thus constructed.

This is not to say that these data have nothing to do with the model. The model posits a certain number of particles and informs us about the way in which they interact, both with each other and with their environment. Using this knowledge, we can place them in a certain experimental context. The data we then gather in an experiment are the product of the elements represented in the model and of the way in which they operate in that context. Characteristically this context is one which we are able to control and about which we have reliable knowledge (e.g. knowledge about detectors, accelerators, photographic plates, and so on). Using this and a model we can derive predictions about what the outcomes of an experiment will be. But, and this is the salient point, these predictions involve the entire experimental set-up and not only the model, and there is nothing in the model itself to which one

---

[38] The model we are talking about here is not the so-called standard model of elementary particles as a whole. Rather, what we have in mind is one specific model about the interaction of certain particles of the kind one would find in a theoretical paper on this experiment.

could compare the data. Hence, data are highly contextual and there is a big gap between observable outcomes of experiments and anything one might call a substructure of a carrier of neutral currents.

To underwrite this claim Bogen and Woodward notice that parallel to the research at CERN, the NAL in Chicago also performed an experiment to detect weak neutral currents, but the data obtained in that experiment were quite different. They consisted of records of patterns of discharge in electronic particle detectors. Although the experiments at CERN and at NAL were totally different and as a consequence the data gathered had nothing in common, they were meant to provide evidence for the same theoretical model. But the model, to reiterate the point, does not contain any of these contextual factors. It posits certain particles and their interaction with other particles, not how detectors work or what readings they show. That is, the model is not idiosyncratic to a special experimental context in the way the data are, and therefore it is not surprising that they do not contain a substructure that is isomorphic to the data. For this reason models, and carriers more generally, represent phenomena, not data.

It is difficult to give a general characterisation of phenomena because they do not belong to one of the traditional ontological categories (Bogen and Woodward 1988, p. 321). In fact, phenomena fall into many different established categories, including objects, features, events, processes, states, and states of affairs, or they defy classification in these terms altogether. This, however, does not detract from the usefulness of the concept of a phenomenon because specifying one particular ontological category to which all phenomena belong is inessential for the purpose of this section. What matters to the problem at hand is the distinctive role they play in connection with representation.

What, then, is the significance of data if they are not the kind of things that carriers represent? The answer to this question is that data perform an evidential function. That is, data play the role of evidence for the presence of certain phenomena. In other words, they provide evidence for the accuracy or correctness of a representation. The fact that we find a certain pattern in a bubble chamber photograph is evidence for the existence of the neutral currents that are represented in the model. This does not denigrate the importance of data in science, but it frees the investigation from the claim that data have to be embeddable into the carrier at stake.

Those who want to establish data models as targets can reply to this in three ways. The first reply is an appeal to radical empiricism. By postulating phenomena over and above data we leave the firm ground of observable things and start engaging in trans-empirical speculation. But science has to restrict its claims to observables and remain silent (or at least agnostic) about the rest. Therefore, so the objection goes, phenomena are chimeras that cannot be part of any serious account of science. It is, however, doubtful that this helps the data model theorist. Firstly, note that it even rules out representing "observable phenomena". To borrow van Fraassen's example, on this story, a population model of deer reproduction would represent data, rather than deer (2008, pp. 254–57). Traditionally, empiricists would readily accept that deer, and the rates at which they reproduce, are observable phenomena. Denying that they are represented, by replacing them with data models,

4.5 The Structure of Target Systems

seems to be an implausible move. Secondly, irrespective of whether one understands phenomena realistically (Bogen and Woodward 1988) or antirealistically (McAllister 1997), it is phenomena that carriers portray and not data. To deny the reality of phenomena just won't make a theoretical model *represent* data. Whether we regard neutral currents as real or not, it is neutral currents that are portrayed in a field-theoretical model, not bubble chamber photographs. Of course, one can suspend belief about the reality of these currents, but that is a different matter.

The second reply is to invoke a chain of representational relationships. Brading and Landry (2006) point out that the connection between a carrier and the world can be broken down in two parts: the connection between a carrier and a data model, and the connection between a data model and the world (2006, p. 575). So the structuralist could claim that carriers represent data models in virtue of an isomorphism between the two and additionally claim that data models in turn represent phenomena. But the key questions that need to be addressed here are (a) what establishes the representational relationship between data models and phenomena, and (b) why if a carrier represented some data model, which in turn represented some phenomenon, would that establish a representational relationship between the carrier and the phenomenon itself? With respect to the first question, Brading and Landry argue that it cannot be captured within the structuralist framework (2006, p. 575). The question has just been pushed back: rather than asking how a scientific model *qua* mathematical structure represents a phenomenon, we now ask how a data model *qua* mathematical structure represents a phenomenon. With respect to the second question, although representation is not intransitive, it is not transitive either (Frigg 2002, pp. 11–12). So if we cannot rely on transitivity of representation in general, then in this context, at the very least more needs to be said regarding how a carrier representing a data model, which in turn represents the phenomenon from which data are extracted, establishes a representational relationship between the first and last element in the representational chain.

The third reply is due to van Fraassen (2006, 2008). His "Wittgensteinian" solution is to diffuse the loss of reality objection. Once we pay sufficient attention to the pragmatic features of the contexts in which carriers and data models are used, van Fraassen claims, there actually is no difference between representing data and representing a target (or a phenomenon in Bogen and Woodward's sense): "in a context in which a given [data] model is *someone's* representation of a phenomenon, there is *for that person* no difference between the question *whether a theory* [theoretical model] *fits that representation* and the question *whether that theory fits the phenomenon*" (2008, p. 259 original italics, underlines omitted). van Fraassen's argument for this claim is long and difficult and we cannot fully investigate it here; we restrict attention to one crucial ingredient, namely Moore's paradox, and refer the reader to Nguyen's (2016) for a detailed discussion of the argument.

Moore's paradox is that we cannot assert sentences of the form "p and I don't believe that p", where p is an arbitrary proposition. For instance, someone cannot assert that Napoleon was defeated in the battle of Waterloo and assert, at the same time, that she doesn't believe that Napoleon was defeated in the battle of Waterloo. van Fraassen's treatment of Moore's paradox is that speakers cannot assert such sentences because the pragmatic commitments incurred by asserting the first conjunct include that the speaker believes that p. This commitment is then contradicted by the assertion of the second conjunct. So instances of Moore's paradox are pragmatic contradictions. van Fraassen then draws an analogy between this paradox and the scientific representation. He submits that a user simply cannot, on pain of pragmatic contradiction, assert that a data model of a target system be embeddable within a carrier without thereby accepting that the carrier represents the target.

However, Nguyen (2016) argues that in the case of using a data model as a representation of a phenomenon, no such pragmatic commitment is incurred, and therefore no such contradiction follows when accompanied by doubt that the carrier also represents the phenomenon. To see why this is the case, consider a more mundane example of representation: a caricaturist can represent Margaret Thatcher as draconian without thereby committing himself to the belief that Margaret Thatcher really is draconian. The caricaturist could, for instance, have been paid by a newspaper to produce a caricature representing Thatcher as being draconian and then simply do his job, but at the same time be privately convinced that Thatcher was not draconian. Or returning to van Fraassen's example of measuring the deer population, the scientist may be sceptical about the techniques used to construct the data model: perhaps camera traps were used to count deer, and it's known that these traps can exhibit malfunction in foggy conditions and heavy rain. In such an instance the scientist can use the data model to represent the actual deer population, without thereby committing herself to the idea that it is accurate, and therefore she can accept that the data model be embedded in a carrier, without thereby committing herself to the belief that the carrier is accurate (in fact, in this instance she could explicitly think the carrier mistaken). The point is that, pragmatically speaking, acts of representation are weaker than acts of assertion: they do not incur the doxastic commitments required for van Fraassen's analogy to go through.

So it seems that van Fraassen doesn't succeed in dispelling the loss of reality objection. How target systems enter the picture in the structuralist account of scientific representation remains therefore a question that structuralists who invoke data models as providing the target end structures must address. Without such an account the structuralist account of representation remains at the level of data, a position that seems implausible, and contrary to scientific practice.

We now turn to the second response: that a structure is instantiated in the system. As mentioned previously, this response comes in three versions. The first is metaphysically more parsimonious and builds on the systems' constituents. Although target systems are not structures, they are composed of parts that instantiate physical properties and relations. The parts can be used to define the domain of individuals, and by considering the physical properties and relations purely extensionally, we arrive at a class of extensional relations defined over that domain.[39] This supplies the

---

[39] This way of introducing a structure can be found, for instance, in Suppes' discussion of the solar system (2002, p. 22).

4.5 The Structure of Target Systems

required notion of structure. We might then say that physical systems instantiate a certain structure, and it is this structure that models are isomorphic to.

As an example, consider the methane molecule. The molecule consists of a carbon atom and four hydrogen atoms grouped around it, forming a tetrahedron. Between each hydrogen atom and the carbon atom there is a covalent bond. One can then regard the atoms as objects and the bonds are relations. Denoting the carbon atom by a, and the four hydrogen atoms by b, c, d, and e, we obtain a structure S with the domain $U = \{a, b, c, d, e\}$ and the relation $r = \{\langle a, b\rangle, \langle b, a\rangle, \langle a, c\rangle, \langle c, a\rangle, \langle a, d\rangle, \langle d, a\rangle, \langle a, e\rangle, \langle e, a\rangle\}$, which can be interpreted as *being connected by a covalent bond*.

The main problem facing this approach is the underdetermination of the target end structure. Underdetermination threatens in two distinct ways, through what we call *the argument from multiple specifications* and through what is generally known as *Newman's theorem*. Let us consider multiple specifications first. In order to identify the structure of a target system we have to do two things: (i) specify a domain of objects and (ii) specify the relevant physical relations that they instantiate. Turning first to (ii), it is worth pointing out that in different contexts, we might specify different physical relations, depending on the purposes of inquiry. And in most contexts of enquiry, we're not interested in *every* physical feature the system has. In the case of the structure of the methane molecule above, we have specified a structure with only one two-place relation, and we haven't specified (in the structure) that it is a which is the carbon atom and $b, c, d,$ and e which are the hydrogen atoms (of course we could have done, which would have involved adding two one-place relations to the structure). So each specification works to select the relevant physical features, and it's only relative to this specification that we can even begin to talk about what "the" structure of the system is.

As regards (i), what counts as an object in a given target system is a substantial question (Frigg 2006). In the case of the methane molecule, one could just as well choose bonds as objects and consider the relation *sharing a node with another bond*. Denoting the bonds by a', b', c', and d', we obtain a structure S' with the domain $U' = \{a', b', c', d'\}$ and the relation $r' = \{\langle a', b'\rangle, \langle b', a'\rangle, \langle a', c'\rangle, \langle c', a'\rangle, \langle a', d'\rangle, \langle d', a'\rangle, \langle b', c'\rangle, \langle c', b'\rangle, \langle b', d'\rangle, \langle d', b'\rangle, \langle c', d'\rangle, \langle d', c'\rangle\}$. Obviously, S and S' are not isomorphic. So which structure is picked out depends on how the system is specified. Depending on which parts one regards as individuals and what relation one chooses, very different structures can emerge. And it takes little ingenuity to come up with further specifications of the methane molecule, which lead to yet other structures.

There is nothing special about the methane molecule, and any target system can be physically specified in different ways, which ground different structures. So the lesson learned generalises: there is no such thing as *the* structure of a target system. Systems only have a structure relative to a specification both of the relevant physical features and of certain parts of the system as the objects which instantiate those features.

Pincock, in discussing a similar problem, argues that the target end structure comes from:

a concrete system along with a specification of the relevant physical properties. This specification fixes an associated structure. Following Suárez, we can say that the system instantiates that structure, relative to that specification, and allow that structural relations are preserved by this instantiation relation [reference omitted]. This allows us to say that a structural relation obtains between a concrete system and an abstract structure. (2012, pp. 28–29)

So the claim is that the identification of "relevant physical properties" solves the problem of underdetermination because it "fixes an associated structure". Pincock's suggestion is a step in the right direction, and we agree that we can only talk about "the" structure of a target system relative to a specification of which physical features are of interest in the modelling context. However, as we have just seen, even if we specify the relevant physical features of the methane molecule by identifying carbon atoms, hydrogen atoms, and covalent bonds, there are still multiple candidate structures compatible with these properties, and these properties do not single out a particular one of these structures as "the" target structure. The problem is that in order to identify a unique target structure, a specification has to do more than just isolate the relevant physical features of the target; it also has to specify them in terms of making up a domain of a structure and of making up the (physical counterparts of) relations defined extensionally over this domain.

In our (2017) publication we provide an alternative way of thinking about how to specify the relevant structure, which relies on the idea that we can *describe* a target system in multiple ways. As before, systems only have a structure under a particular description and there are many non-equivalent descriptions. We suggest that in general such a description, $\boldsymbol{D_S}$, has the following form: the target system T contains objects $\boldsymbol{o_1}, \ldots, \boldsymbol{o_n}$ that either individually, or in collections, instantiate physical properties and relations $\boldsymbol{r_1}, \ldots, \boldsymbol{r_m}$. The use of bold font indicates that the terms in $\boldsymbol{D_S}$ refer to physical properties and relations: $\boldsymbol{D_S}$ contains terms that describe the system in physical, not set-theoretic terms. The terms individuating the $\boldsymbol{o}$'s are terms like "covalent bond" or "carbon atom" and not "element of a set". Likewise, the terms individuating the $\boldsymbol{r}$'s are terms like "is bonded to" or "shares a node with" and not "being an n-tuple". In the previous example the first description is $\boldsymbol{D}$: "the methane molecule contains a carbon atom, $\boldsymbol{a}$; a hydrogen atom, $\boldsymbol{b}$; a hydrogen atom, $\boldsymbol{c}$; and a hydrogen atom, $\boldsymbol{d}$ where the carbon atom is bonded, $\boldsymbol{r}$, to each of the hydrogen atoms".

We call $\boldsymbol{D_S}$ a *structure generating description* because it generates a structure when we remove the physical nature of the objects and relations from $\boldsymbol{D_S}$. In the case of the $\boldsymbol{o}$'s in $\boldsymbol{D}$ for example, this means that we move from describing the system as consisting of atoms to describing it as consisting of objects. When $\boldsymbol{D}$ talks of four atoms, we now remove "atoms" from this description and only keep "objects" (and, possibly, a numbering of them as the first object, the second object, etc.). The relations get stripped of their physical nature by replacing them with their extension. $\boldsymbol{D}$ describes the relation as "being bonded to". We now remove "being bonded to" and only keep the extension of the relation, which only specifies between which objects the relation holds, but not what the relation itself is. In general, then, this procedure allows us to pass from a description $\boldsymbol{D_S}$ containing physical terminology to a description D_S (now no longer in bold font), which is purely structural. While $\boldsymbol{D_S}$ describes

4.5 The Structure of Target Systems

T as consisting of atoms sharing covalent bonds, D_S describes it as consisting of so-and-so many objects that enter into purely extensionally defined relations. D_S is a structural description, which specifies a structure consisting of n dummy-objects $o_1, ..., o_n$ which enter into purely extensionally defined relations $r_1, ..., r_m$ (note that symbols are now no longer in bold font). This provides the target end structure required by the structuralist.[40]

It is obvious that different structure generating descriptions of the same system can be given. Rather describing the methane molecule through $\boldsymbol{D}$, we can describe it through $\boldsymbol{D'}$: "the methane molecule contains a bond, $\boldsymbol{a'}$; a bond, $\boldsymbol{b'}$; a bond, $\boldsymbol{c'}$; and a bond, $\boldsymbol{d'}$ where each bond shares a node, $\boldsymbol{r'}$, with each other bond". Again, the description contains terms that refer to physical objects, properties, and relations. Abstracting away from the concrete concepts yields D', which yields the second structure mentioned previously.

It's important to note that *both* of the structure generating descriptions can equally legitimately be applied to the methane molecule, and yet both generate a different structure. As such, on this approach, "the" structure of a target system is always relative to a particular way of describing it in physical terms.[41] This renders talk about a carrier being isomorphic to the target system, or even *the* target system's structure *simpliciter*, meaningless. Structural claims do not "stand on their own" in that their truth rests on the truth of a more concrete description of the target system. As a consequence, structure generating descriptions, which are stated in physical vocabulary, are an integral part of an analysis of epistemic representation.

How much of a problem this is depends on how austere one's conception of structuralism is. The semantic view of theories, in which the structuralist notion of models originates, was in many ways the result of an anti-linguistic turn in the philosophy of science. Many proponents of the view aimed to exorcise language from an analysis of theories, and they emphasised that the model-world relationship ought to be understood as a *purely* structural relation.[42] For structuralists of that stripe the above argument is bad news. However, a more attenuated position could integrate descriptions in the package of modelling, but this would involve abandoning

---

[40] It's worth mentioning that in our (2017) we consider the question of mathematical applicability under a broadly structuralist understanding about mathematics rather than a structuralist account of representation per se. So even if the approach of generating target end structures via physical descriptions is successful, this doesn't vindicate structuralist accounts of representation given the other issues, discussed in this chapter, that such accounts face. See Chap. 9 for a further discussion of this approach to the Applicability of Mathematics Condition outside of the structuralist account of epistemic representation.

[41] Frigg (2006, pp. 55–56) provides another argument that pulls in the same direction: structural claims are abstract and are true only relative to a more concrete non-structural description. For a critical discussion of this argument see Frisch's (2015, pp. 289–94) and Portides' (2017, pp. 43-44).

[42] van Fraassen, for instance, submits that "no concept which is essentially language dependent has any philosophical importance at all" (1980, p. 56) and observes that "[t]he semantic view of theories makes language largely irrelevant" (1989, p. 222). Other proponents of the view, while less vocal about the irrelevance of language, have not assigned language a systematic place in their analysis of theories.

the idea that representation can be cashed out solely in structural terms. Bueno and French have recently endorsed such a position. They accept the point that different descriptions lead to different structures and explain that such descriptions would involve "at the very least some minimal mathematics and certain physical assumptions" (Bueno and French 2011, p. 887). Likewise, "Munich" structuralists explicitly acknowledge the need for a concrete description of the target system (Balzer et al. 1987, pp. 37–38), and they consider these "informal descriptions" to be "internal" to the theory. This is a plausible move, but those endorsing this solution have to concede that there is more to epistemic representation than just structures and morphisms.

As noted previously, there is a second way in which structural indeterminacy can surface, namely via Newman's theorem. The theorem essentially says that any system instantiates any structure, the only constraint being cardinality.[43] The details of the theorem go beyond our current purposes, but it is illustrative to sketch how it works.[44]

Let $S_X = \langle U, R \rangle$ be a carrier structure such that the cardinality $|U|$ of U is k. Let $T = \{x_1, \ldots, x_k\}$ be a set consisting of the k objects from the target system. Since $|T| = |U| = k$, there is a bijection $f: T \rightarrow U$. Using this bijection, for each $r_i \in R$, we can construct a set r'_i consisting of n-tuples of objects from T as follows: $\{\langle x_i, \ldots, x_j \rangle : \langle f(x_i), \ldots, f(x_j) \rangle \in r_i\}$. Collecting these relations together gives a structure, $\langle T, R' \rangle$ which is isomorphic to $\langle U, R \rangle$ by construction. Hence, *any* structure of cardinality k is isomorphic to a target of cardinality k because the target instantiates any structure of cardinality k.

This result makes isomorphism too cheap to come by. At the root of the problem is the fact that each relation $r'_i \in R'$ has been constructed simply by collecting elements of the target system into n-tuples based solely on the relations defined over elements in the image of f, without any heed to paid to the question of whether the elements of T are related in any significant way. This is possible because, as we have seen previously, the relations of a structure are defined purely extensionally. What defines a relation is not its intension, but the set of n-tuples between which it holds, and this allows for relations to be constructed in the way required by Newman's theorem. This move can be blocked (thereby resolving problem that isomorphism is too cheap), but any move that blocks Newman's theorem requires that among all structures formally instantiated by a target system one is singled out as being the true or natural structure of the system. How to do this in the structuralist tradition remains a matter of controversy.[45]

Newman's theorem is both stronger and weaker than the argument from multiple specifications. It's stronger in that it provides more alternative structures than multiple specifications. It's weaker in that many of the structures it provides are

---

[43] An analogous conclusion is reached in Putnam's so-called model-theoretic argument. See Demopoulos' (2003) for a discussion.

[44] For more detailed discussions see Frigg and Votsis' (2011) and Ketland's (2004).

[45] Ainsworth's (2009) provides as useful summary of different solutions.

4.5 The Structure of Target Systems

"unphysical" because they are purely set-theoretic combinations of elements. Any collection of n objects in the target system can be collected together into an n-tuple. And any collection of n-tuples can be collected together into an n-place relation. So the associations between the objects under the construction of Newman's theorem can be arbitrary. By contrast, as long as the language of the structure generating description is an ordinary physical language, such descriptions pick out structures that a system can reasonably been seen as possessing.

This completes our discussion of the first version of the structure instantiation account, and we now turn to the second version. This version also emerges from the literature on the applicability of mathematics. Structural Platonists like Resnik (1997) and Shapiro (1983, 1997, 2000) take structures to be "ante rem" universals. On this view, structures exist independently of physical systems, yet they can be *instantiated* in physical systems.[46] The instantiation relation that holds between a target system and a structure is supposed to be the sort of relation that holds between an object and a property (the phone box instantiating redness), or a pair of objects and a relation they enter into (Hooke and Newton having been involved in a vindictive rivalry), and so on. This comes across most clearly when structuralists in the philosophy of mathematics discuss the relationship between mathematics and the physical world. For example, Shapiro argues that:

> On [the structuralist account of mathematics] the problem of the relationship between mathematics and reality is a special case of the problem of the instantiation of universals. Mathematics is to reality as universal is to instantiated particular. As above, the "universal" here refers to a pattern or structure; the "particular" refers not to an individual object, but to a system of related objects. More specifically, then, mathematics is to reality as pattern is to patterned. (1983, p. 538)[47]

This view raises all kind of metaphysical issues about the ontology of structures and the instantiation relation. Let us set aside these issues and assume that they can be resolved in one way or another.[48] This, however, still leaves us with serious epistemic and semantic questions. How do we know that a certain structure is instantiated in a system and how do we refer to it? Objects do not come with labels on their sleeves specifying which structures they instantiate, and proponents of structural universals face a serious problem in providing an account of *how we access* the structures instantiated by target systems. Even if – as a brute metaphysical fact – target systems only instantiate a small number of structures (and Newman's theorem

---

[46] An alternative to this Platonist kind of structuralism is Hellman's modal structuralism (1989, 1996), which construes infinite domains as modal constructs.

[47] A similar position is found in Resnik (1997, p. 204). Due to the fact that some mathematical structures are not instantiated by any physical system, Shapiro later considers whether or not structures as universals should be thought of as ante rem universals, as in re universals (eliminative structuralism), or in modal terms (Shapiro 1997). These distinctions are immaterial to our question. But, a word of warning about terminology is in order here. Shapiro often uses the term "exemplify" to refer to the relationship between a universal and a physical system. We prefer "instantiate" because, as we will see in Chaps. 7, 8, and 9, following Goodman "exemplification" is used in a slightly different way in the literature on representation.

[48] For a relevant discussion in the context of universals see Armstrong's (1989).

is assumed to have no bite here), we still don't know how we can ever come to know whether or not a certain structure is in fact instantiated in the target. It seems that individuating a domain of objects and identifying relations between them is the only way for us to access a structure. But then we are back to the first version of the response, and we are again faced with all the problems that relying on descriptions raises.

The third version of the the view that targets "have" structures is radical. One might take target systems themselves to *be* structures. On this view, targets, first appearances notwithstanding, are not ordinary physical systems; they literally are set-theoretic structures. So all of the objects in a system would be considered purely as "dummy" objects, being individuated only by their "objecthood" (or even more extremely, only by the relations, construed purely extensionally, they have with other objects). And all of the properties and relations they have would be considered purely extensionally: with only the logical properties of the relations mattering. Thus, it would be of no importance whether a relation was "being an ancestor of" or "being to the left of" as long as they picked out the same *n*-tuples of objects.

If this is the case, then there is no problem after all with the idea that carriers can be isomorphic to a target. Carrier structures are then isomorphic to target structures, just as we assumed at the beginning of Sect. 4.3. One might expect ontic structural realists to take this position. If the world fundamentally is a structure, then there is nothing mysterious about the notion of an isomorphism between a carrier and the world. Surprisingly, some ontic structuralists have been hesitant to adopt such a view (see French and Ladyman's (1999, p. 113) and French's (2014, p. 195)). Others, however, seem to endorse it. Tegmark (2008), for instance, offers an explicit defence of the idea that the world simply is a mathematical structure. He defines a seemingly moderate form of realism – what he calls the "external reality hypothesis (ERH)" – as the claim that "there exists an external physical reality completely independent of us humans" (ibid., p. 102) and argues that this entails that the world is a mathematical structure (his "mathematical universe hypothesis") (ibid.). His argument for this is based on the idea that a so-called "theory-of-everything" must be expressible in a form that is devoid of human-centric "baggage" (by the ERH), and the only theories that are devoid of such baggage are mathematical, which, strictly speaking, describe mathematical structures. Thus, since a complete theory-of-everything describes an external reality independent of humans, and since it describes a mathematical structure, the external reality itself *is* a mathematical structure.

Butterfield (2014) argues that the argument is invalid because it equivocates on the meaning of "is" in "the world is a mathematical structure". There are at least two ways of using the word "is". The first is predication, for example when we say "the apple is red". The second is identity, for example when we say "Le Corbusier is Charles-Edouard Jeanneret". Butterfield submits that Tegmark's argument will deliver a conclusion according to which the world is a mathematical structure in the first but not the second sense. Tegmark, however, seems to read the claim in the second sense. In other words, even if one accepts that the ERH might deliver the conclusion that the world instantiates a mathematical structure, it doesn't deliver the

4.5 The Structure of Target Systems

stronger result that the world is a structure in the sense of an identity claim. But if the argument only delivers the conclusion that the world instantiates a certain structure, we're back to the second version of the response.

Even if one were willing to set concerns about "is" aside, Tegmark's approach stands or falls on the strengths of the premise that a complete theory of everything will be formulated purely mathematically, without any "human baggage", which in turn relies on a strict reductionist account of scientific knowledge (ibid., pp. 103–04). Discussing this in any detail goes beyond our current purposes. But it is worth noting that Tegmark's discussion is focused on the claim that *fundamentally* the world is a mathematical structure. Even if this were the case, it seems irrelevant for many of our current scientific representations, whose targets aren't at this level, for two related reasons. First, for those sorts of representations, their targets are not described in fundamental terms. When modelling an aeroplane wing we don't refer to the fundamental super-string structure of the bits of matter that make up the wing (or whatever else current "fundamental" physics takes to be the basic building blocks of matter), and we don't construct wing models that are isomorphic to such fundamental structures. So even if the world is fundamentally a structure, this structure is not prima facie the target of many of our models. Given this, in order for Tegmark's argument to go through, any of the considerations, which (apparently) yield the conclusion that at a fundamental level the world is a mathematical structure, would have to hold at higher levels too, thereby ensuring that the world is a structure at all levels. And it's far from obvious that this would be the case, even assuming a reductionist account of the relationships between these levels. So Tegmark's account offers no answer to the question about where structures are to be found at the level of non-fundamental target systems.

In sum, then, the structuralist view of representation faces a unique difficulty concerning where the target end structure comes from. In a sense this is tied up with broader questions concerning the metaphysics or ontology of the world: questions that in general one might hope to be independent of questions concerning epistemic representation. However, in virtue of tying their view of epistemic representation up with a morphism between a carrier and the structure of the target system, the structuralist view of representation is therefore committed to providing the metaphysical/ontological details of where the latter comes from. The problem is even more pressing for those who don't demarcate beyond distinguishing between direct and indirect representations: not only do they have to provide an account of target end structure; they also have to provide an account of where the structure of the carrier comes from in cases where the carriers are not obviously thought of in mathematical or structural terms (as is the case with material models, and epistemic representations including the likes of works of art and maps).

Chapter 5
The Inferential View

In this chapter we discuss accounts of scientific representation that analyse representation in terms of the inferential role that models play in scientific investigations. According to the accounts discussed earlier, a model's inferential capacity – its capacity to ground surrogative reasoning – dropped out of an analysis of epistemic representation: namely proposed morphisms or hypothesised similarity relations between models and their targets. The core idea of inferentialism is to analyse scientific representation directly in terms of a model's inferential capacities without a "detour" through morphisms or similarity relations: a model represents its target if the model's users can draw inferences about the target from the model. So inferentialism meets the *Surrogative Reasoning Condition* by construction, as it were.

We begin by discussing where inferentialism stands on the issues concerning demarcation, and we will see that the common versions of inferentialism give negative answers to both demarcation problems and therefore end up addressing the ER-Problem (Sect. 5.1). We discuss Suárez deflationary answer to the problem (Sect. 5.2) and then consider the motivations behind a deflationary account (Sect. 5.3). We continue to consider first Contessa's, then Díez's, and finally Bolinska's and Ducheyne's reinflated versions of inferentialism (Sect. 5.4).[1]

5.1 The Demarcation Problems

According to the inferential view, what makes something a model is that it licences inferences about its target system. But nothing in the view entails that this is restricted to models or scientific representations more generally. For this reason,

---

[1] In our (2017a) we followed Suárez (2015) in classifying Hughes's DDI account as an inferential account, or a close cousin thereof. However, from a systematic point of view, the DDI account is better classified as an analysis of *representation-as*, and so we now discuss it in Chap. 7.

advocates of the view give negative answers to both demarcation problems. This is fully explicit with respect to the Scientific Representational Demarcation Problem. Suárez (2004) argues that:

> if the inferential conception is right, scientific representation is in several respects very close to iconic modes of representation, like painting. Representational paintings, such as Velázquez's portrait of Innocent X permit us to draw inferences regarding those objects that they represent. The Velázquez canvas allows us to infer some personal qualities of the Pope and some of the properties of the Catholic Church as a social institution, as well as the Pope's physical appearance. (p. 777)

He also emphasises that an "important virtue of the inferential conception is its ability to capture the representational or nonrepresentational distinction in art as well as science" (ibid.).[2] Contessa (2007) is explicit that the target of his analysis is "epistemic representation" in the sense introduced in Chap. 1:

> Portraits, photographs, maps, graphs, and a large number of other representational devices usually allow their users to perform (valid) inferences to their targets and, as such, according to [that] characterization […] would be considered epistemic representations of their targets (for us). For example, according to that characterization, if we are able to perform (valid) inferences from a portrait to its subject (as we usually seem able to do), then the portrait is an epistemic representation of its subject (for us). (p. 54)

With respect to the Taxonomic Representational Demarcation Problem things are a little more open. However, both Suárez and Contessa illustrate their accounts with a large variety of representational types, which implies that they do not demarcate taxonomically. Suárez (2004, p. 774) also discusses his account as applying to linguistic entities such as a quantum state diffusion equation, which suggests that he also does not subscribe to the distinction between direct and indirect forms of representation as discussed in Sect. 3.1. We take this as evidence that inferentialists don't demarcate taxonomically and hence analyse inferential views as attempts to address the ER-Problem: in virtue of what is X an epistemic representation of T?

---

[2] Occasionally Suárez suggests that the inferential capacity condition is supposed to distinguish scientific from non-scientific representation e.g. "the inferential conception adds a second condition, which is specifically required for *scientific* representation" (2010, p. 98 original emphasis) but since this condition also applies to non-scientific representations, we take him to be concerned with distinguishing epistemic from and non-epistemic representations rather than with demarcating scientific from non-scientific epistemic representation.

5.2 Deflationary Inferentialism

Suárez argues that we should adopt a "deflationary or minimalist attitude and strategy" (2004, p. 770) when addressing the ER-Problem. We will discuss deflationism in some detail in the next section, but in order to formulate and understand Suárez's theory of representation we need at least a preliminary idea of what is meant by a deflationary attitude. In fact, two different notions of deflationism are in operation in his account. The first is "abandoning the aim of a substantive theory to seek universal necessary and sufficient conditions that are met in each and every concrete real instance of scientific representation [...] necessary conditions will certainly be good enough" (ibid., p. 771). We call the view that a theory of epistemic representation should provide only necessary conditions n-deflationism ("n" for "necessary"). The second notion is that we should seek "no deeper features to representation other than its surface features" (ibid.) or "platitudes" (Suárez and Solé 2006, p. 40), and that we should deny that an analysis of a concept "is the kind of analysis that will shed explanatory light on our use of the concept" (Suárez 2015, p. 39). We call this position s-deflationism ("s" for "surface feature"). As far as we can tell, Suárez intends his account of epistemic representation to be deflationary in both senses.

Suárez dubs the account that satisfies these criteria "inferentialism" and provides the following characterisation of epistemic representations (Suárez 2004, p. 773):[3]

Inferentialism 1: a carrier X is an epistemic representation of a target T only if (i) the representational force of X points towards T, and (ii) X allows competent and informed agents to draw specific inferences regarding T.

Notice that this view is not an instance of the ER-Scheme: in keeping with n-deflationism it features a material conditional rather than a biconditional and hence provides necessary (but not sufficient) conditions for X to represent T. We now discuss each condition in turn, trying to explicate in what way they satisfy s-deflationism.

The first condition is designed to ensure that X and T enter into a representational relationship, and Suárez stresses that representational force is "necessary for any kind of representation" (ibid., p. 776). But explaining representation in terms of representational force seems to shed little light on the matter as long as no analysis of representational force is offered. Suárez addresses this point by submitting that the first condition can be "satisfied by mere stipulation of a target for any source" (ibid., p. 771), where he uses "source" to refer to what we call "carrier" (see Sect. 1.1). This might look like denotation as discussed in Chaps. 2 and 3 (as well as Chaps. 7–9 later in the book), but Suárez stresses that this is not what he intends for

---

[3] A similar idea has been introduced by Bailer-Jones who says "I call the relationship between models and propositions entailment: models entail propositions" (2003, p. 60). However, she immediately adds that "interpreting models in terms of the propositions entailed by them can be only part of the story regarding the representational relationship between models and the empirical world" (ibid., p. 61), which runs counter to the inferentialist doctrine that representation has to be accounted for solely, or at least primarily, in terms of inferences.

two reasons. Firstly, he takes denotation to be a substantive relation between a carrier and its target, and the introduction of such a relation would violate the requirement of s-deflationism (Suárez 2015, p. 41). Secondly, X can denote T only if T exists. Thus, including denotation as a necessary condition on scientific representation, "would rule out fictional representation, that is, representation of nonexisting entities" (Suárez 2004, p. 772), but "any adequate account of scientific representation must accommodate representations with fictional or imaginary targets" (Suárez 2015, p. 44).

The latter issue is one that besets other accounts of representation too, in particular similarity and isomorphism accounts. In Chap. 7 we encounter a solution (due to Goodman and Elgin), which keeps denotation as a condition without running up against this difficulty.[4] So non-existent targets need not necessarily be a reason to ban denotation from a theory of representation. The former issue, however, goes right to the heart of Suárez's account: it makes good on the s-deflationary condition that nothing other than surface features can be included in an account of representation. At a surface level one cannot explicate "representational force" and any attempt to specify what representational force consists in is a violation of s-deflationism.

The second necessary condition, that carriers allow competent and informed agents to draw specific inferences about their targets, is in fact the Surrogative Reasoning Condition that we introduced in Chap. 1, now taken as a necessary condition on epistemic representation. The sorts of inferences that carriers allow are not constrained. Suárez points out that the condition "does not require that [X] allow deductive reasoning and inference; any type of reasoning – inductive, analogical, abductive – is in principle allowed" (Suárez 2004, p. 773).

A problem for this approach is that we are left with no account of how these inferences are generated: what is it about carriers that allows them to licence inferences about their targets, or what leads them to licence some inferences and not others? Contessa makes this point most stridently when he argues that:

> On the inferential conception, the user's ability to perform inferences from a vehicle to a target seems to be a brute fact, which has no deeper explanation. This makes the connection between epistemic representation and valid surrogative reasoning needlessly obscure and the performance of valid surrogative inferences an activity as mysterious and unfathomable as soothsaying or divination. (2007, p. 61)

This seems correct, but Suárez can dismiss this complaint by appeal to s-deflationism. Since inferential capacity is supposed to be a surface level feature of scientific representation, we are not supposed to ask for an elucidation about what makes an agent competent and well informed and how inferences are drawn.

Suárez's position provides us with a concept of epistemic representation that is cashed out in terms of an inexplicable notion of representational force and of an equally inexplicable capacity to licence inferences. This seems to be very little

---

[4] Suárez discusses but doesn't endorse Goodman and Elgin's way of dealing with denotation; see his (2010, p. 97) and (2015, p. 44).

indeed. It is the adoption of a deflationary attitude that allows him to block any attempt to further unpack these conditions and so the crucial question is: why should one adopt deflationism (of either stripe)?

We turn to this question in the next section. Before doing so, we want to briefly outline how the above account fares with respect to the other problems introduced in Chap. 1. The account provides a neat explanation of the possibility of misrepresentation. As Suárez points out, condition (ii) of Inferentialism 1 "accounts for inaccuracy since it demands that we correctly draw inferences from the source about the target, but it does not demand that the conclusions of these inferences be all true, nor that all truths about the target may be inferred" (2004, p. 776). Carriers represent their targets only if they license inferences about them; they represent them accurately to the extent that the conclusions of these inferences are true. This is how Suárez's account meets the Misrepresentation Condition.

Given the wide variety of types of representation that this account applies to, it's unsurprising that Suárez shows little interest in the Problem of Carriers. The only constraint that Inferentialism 1 places on answers to the Problem of Ontology is that Inferentialism 1 "requires [X] to have the internal structure that allows informed agents to correctly draw inferences about [T]" (Suárez 2004, p. 774). The Problem of Handling is not one that an s-deflationist would want to address, because saying something about how carriers are dealt with would amount to saying how inferences are drawn, thereby infringing deflationist imperatives. And, relatedly, since the account is supposed to apply to a wide variety of entities, including equations and mathematical structures, the account implies that mathematics is successfully applied in the sciences. But in keeping with the spirit of deflationism no explanation is offered about how this is possible.

Suárez does not directly address the Problem of Style, but a minimalist answer emerges from what he says about inferences. On the one hand, he explicitly acknowledges that many different kinds of inferences are allowed by the second condition in Inferentialism 1. In the passage quoted above he mentions inductive, analogical, and abductive inferences. This could be interpreted as the beginning of classification of representational styles. On the other hand, Suárez remains silent about what these kinds are and about how they can be analysed. This is unsurprising because spelling out what these inferences are, and what features of the carrier ground them, would amount to giving a substantial account, which is something Suárez wants to avoid.

5.3 Motivating Deflationism

Let us now return to the question about the motivation for deflationism. As we have seen, a commitment to deflationism about the concept is central to Suárez's approach to epistemic representation. But deflationism comes in different guises, which Suárez illustrates by analogy with deflationism with respect to truth. Thus, in order to understand the motivation for a deflationary view of epistemic representation, it is worth briefly reviewing the motivations for deflationary views of truth.

What it means to provide a "theory of truth" is a significant question in its own right. For our current purposes, what matters is that such a theory is supposed to provide an analysis of sentences of the form "'p' is true", where p is a sentence (or proposition).[5] To use Tarski's classical example: how should we analyse sentences like "'snow is white' is true"? On the face of it, such sentences have the form of attributing a property, *is true*, to an object, the sentence "p". Thought of in these terms, the question then is how we should understand the property *is true*?

A natural reaction to this question is to attempt to analyse what all true sentences have in common: in virtue of what are true sentences true (and false sentences false)? Two prominent traditions of attempts to answer this question are the *correspondence theory of truth* and the *coherence theory of truth*. According to the former, true sentences are true in virtue of "corresponding", in some sense, to what is the case. So the sentence "snow is white" is true because it corresponds to the fact that snow is white; the sentence "grass is green" is true because it corresponds to the fact that grass is green; and so on. Likewise, the sentence "apples are violet" is false because there is no fact to which it corresponds. According to coherentist theories of truth, what makes sentences true or false is not to be explicated in terms of their correspondence to something out there in the world, but rather in terms of their relationships within a wider body of sentences. So what makes it the case that "snow is white" is true is not that it corresponds to snow being white, but rather that the sentence is a member of a set of sentences that mutually support one another. What makes "apples are violet" false is not its failure to correspond to facts in the world, but rather that it fails to cohere with other accepted sentences.

Explicating the details of either kind of theory of truth is far from straightforward, and, at least in part because of this, some theorists began to question the underlying assumption behind these views. In appealing to correspondence or coherence in their analyses of truth, defenders of those views appear to be committed to the idea that a substantial analysis of the truth property can, and should, be given. The analysis is "substantial" in the sense that truth is explicated in terms of another property that sentences can possess, where that property is such that it can be studied independently of the notion of truth, which allows these accounts to give a reductive analysis of truth in terms of that other property.

Advocates of deflationary views of truth deny that this can, or should, be done. Whilst the denial of the need for a substantial analysis of truth is what all deflationary views share, there are different ways to turn this denial into a positive account. Given that our discussion of theories of truth is motivated only by the need to understand the analogy with Suárez's deflationism about epistemic representation, we follow his classification and distinguish between three kinds of deflationary theories of truth (Suárez 2015): the *redundancy* theory (associated with Frank Ramsey and

---

[5] For ease of formulation, we only talk about sentences in what follows. Kirkham's (1992, Chap. 1), Künne's (2003), and Glanzberg's (2018) provide extensive discussions of what it means to provide a theory of truth and survey common accounts.

5.3 Motivating Deflationism

also referred to as the *no theory view*), *abstract minimalism* (associated with Crispin Wright) and the *use theory* (associated with Paul Horwich).[6] According to Suárez, what these accounts have in common is a commitment to what is called the *disquotational schema* – "'p' is true iff p", where p is a variable that ranges over sentences – and that they "either do not provide an analysis [of truth] in terms of necessary and sufficient conditions, or if they do provide such conditions, they claim them to have no explanatory purchase" (ibid., p. 37).

Beginning with the redundancy theory, Suárez characterises the view through the idea that "the terms 'truth' and 'falsity' do not admit a theoretical elucidation or analysis. But that, since they can be eliminated in principle – if not in practice – by disquotation, they do not in fact require such an analysis" (ibid., p. 39). The idea is that the terms can be eliminated because any instance of the form "'p' is true" can be replaced simply by "p" itself. So, as Suarez characterises the position, the redundancy theory denies that any necessary and sufficient conditions for application of the truth property or predicate can be given. He argues that "the generalization of this 'no-theory theory' for any given putative concept X is the thought that X neither possesses nor requires necessary and sufficient conditions because it is not in fact a 'genuine', explanatory or substantive concept" (ibid.).[7]

This approach faces a number of challenges. First, the argument is based on the premise that if deflationism, and in particular the redundancy theory, is good for truth, then it must also be good for epistemic representation. Surprisingly, Suárez offers little by way of explicit argument in favour of any sort of deflationary account of epistemic representation. There is, however, a question whether the analogy between truth and representation is sufficiently robust to justify subjecting them to the same theoretical treatment. In fact, the natural analogue of the linguistic notion of truth is *accurate* epistemic representation, rather than epistemic representation itself, which may be more appropriately compared with linguistic meaning. Second, the argument insinuates that the redundancy theory (or indeed any version of deflationism) is the correct analysis of truth. This, however, is far from an established fact. Different positions are available in the debate and whether deflationism (or any specific version of it) is superior to other proposals remains a matter of controversy.[8] But as long as it's not clear that a redundancy theory about truth is a superior position, it's hard to see how one can muster support for support for deflationism about epistemic representation by appealing to it.

---

[6] We note that there is room to disagree with respect to how to characterize each of these authors, and with respect to how the details of their accounts fit with deflationism. For our current purposes we follow Suarez's explication of the positions, but this shouldn't be taken to imply that we subscribe to his classification and characterisation of different kinds of deflationism, or that the authors in question do so. For useful discussions of deflationary theories of truth see Halbach's (2014) and Stoljar and Damnjanovic's (2014).

[7] Although one might ask why such a position would allow even necessary conditions, Suárez doesn't discuss this.

[8] See the references in footnotes 5 and 6.

Furthermore, a position that allows only necessary conditions on epistemic representation faces a serious problem when it comes to identifying representations. While such an account allows us to *rule out* certain scenarios as instances of epistemic representation (e.g. a proper name doesn't allow a competent and well informed language user to draw inferences about its bearer, and Callender and Cohen's salt-shaker doesn't allow a user to draw inferences about Madagascar), the lack of sufficient conditions doesn't allow us to *rule in* a scenario as an instance of epistemic representation. So on the basis of Inferentialism 1 we are never in position to assert that a particular carrier actually is an epistemic representation!

The other two deflationary positions in the debate over truth are abstract minimalism and the use theory. Suárez characterises the use theory as being based on the idea that "truth is nominally a property, although not a substantive or explanatory one, which is essentially defined by the platitudes of its use of the predicate in practice" (2015, p. 40). Abstract minimalism is presented as the view that while truth is "legitimately a property, which is abstractly characterized by the platitudes, it is a property that cannot explain anything, in particular it fails to explain the norms that govern its very use in practice" (ibid.). Both positions imply that necessary and sufficient conditions for truth *can* be given (ibid.). But on either account, such conditions only capture non-explanatory surface features. This motivates s-deflationism.

Since s-deflationism explicitly allows for necessary and sufficient conditions, Inferentialism 1 can be extended to an instance of the ER-scheme, providing necessary and sufficient conditions:[9]

Inferentialism 2: a carrier X is an epistemic representation of a target T iff (i) the representational force of X points towards T, and (ii) X allows competent and informed agents to draw specific inferences regarding T.

If one takes conditions (i) and (ii) to refer to "features of activates within a normative practice" that "do not stand for relations between sources and targets" (Suárez 2015, p. 46), then we arrive at a "use-based" account of epistemic representation, which Suárez associates with Horwich. In order to understand a particular instance of a carrier X representing a target T we have to understand how agents go about establishing that X's representational force points towards T, as well as the inferential rules they use and the particular inferences draw from X about T.

Plausibly, such a focus on practice amounts to looking at the inferential rules employed in each instance, or type of instance, of epistemic representation. This, however, raises a question about the status of any such analysis vis-à-vis the general theory of representation as given in Inferentialism 2. There seem to be two options: to regard the key terms of Inferentialism 2 as abstract concepts and see a study of representational practice as an inquiry into the concrete basis of these concepts, or to affirm the status of Inferentialism 2 as an exhaustive theory of epistemic representation and embed it into broader inferentialist philosophy of language.

---

[9] Which also seems to be in line with Suárez and Solé (2006, p. 41), who provide a formulation of inferentialism with a biconditional.

5.3 Motivating Deflationism

The first option is to interpret Inferentialism 2 as providing abstract conditions that require concretization in each instance of epistemic representation. Some concepts are more abstract than others. *Game* is more abstract than *chess* or *soccer*; *work* is more abstract than *weeding the garden* or *cleaning the kitchen*; *enjoyment* is more abstract than *seeing a good movie* or *eating chocolate*; and *travelling* is more abstract than *taking a train* or *riding a bicycle*. Intuitively it is clear why this is so. But what is it for one concept to be more abstract than another? Cartwright (1999a, p. 39) provides us with two jointly necessary and sufficient conditions for a concept to be abstract:

> First, a concept that is abstract relative to another more concrete set of descriptions never applies unless one of the more concrete descriptions also applies. These are the descriptions that can be used to "fit out" the abstract description on any given occasion. Second, satisfying the associated concrete description that applies on a particular occasion is what satisfying the abstract description consists in on that occasion.

Consider the example of travelling. The first condition says that unless I either take a train, ride a bicycle, drive a car, or …, then I am not travelling. The second condition says that my riding a bicycle right now is what my travelling consists in. The salient point is that I am only doing one thing, namely riding a bicycle. It is not the case that I am doing two things, riding a bicycle and travelling; I am doing just one thing because riding a bicycle is what my travelling at this particular moment amounts to. To say that I am travelling is just a more abstract description of the very same activity.

If we regard *representational force* and *allowing competent agents to draw inferences regarding T* as abstract concepts of the same kind as *travelling*, then one can see a study of the practices of representation as a study of the concrete realisations of the abstract concepts that feature in Inferentialism 2. The conditions in Inferentialism 2 then provide a blank that has to be filled in every instance of epistemic representation, and they serve as an invitation to take different cases of representation – be they paintings or scientific models – and ask for each of these; first, where their representational force comes from and, second, what makes them cognitively relevant. This approach is plausible, but it renders deflationism otiose. Thus understood, the view becomes indistinguishable from a theory that accepts the Surrogative Reasoning Condition and the Requirement of Directionality as conditions of adequacy and analyses them in pluralist spirit, that is, under the assumption that these conditions can be met in different ways in different contexts, and such a pluralist programme can be carried out without ever mentioning deflationism (as we will see below, it is not obvious that such an approach is still deflationary).

The second option is to stick to inferentialism's original intentions and reaffirm that there is nothing to be said about representation over and above the conditions stated in Inferentialism 2. This, however, would imply that any analysis of the workings of a particular carrier would fall outside the scope of a theory of representation because any attempt to address Contessa's objection would push the investigation outside the territory delineated by s-deflationism. Such an approach seems to be overly purist, and it would leave us with no reason why someone interested in a theory of representation would want to embark in such a study in the first instance.

One might reply this is too bleak a view on the role of practice in a deflationary account of epistemic representation and argue that thinking about it in these terms gets the order of explanation the wrong way around. Rather than attempting to investigate the conditions of epistemic representation by investigating the representational practices that establish it in every instance, one could instead take those conditions as foundational, and investigate how they give rise to representational practices, practices which themselves are explained by the inferentialist's conditions (rather than explaining them). Such an approach is inspired by Brandom's (1994, 2000) inferentialism in the philosophy of language where the central idea is to reverse the order of explanation from representational notions – like truth and reference – to inferential notions – such as the validity of argument. We are urged to begin from the inferential role of sentences (or propositions, or concepts, and so on) – that is, from the role that they play in providing reasons for other sentences (or propositions etc.), and having such reasons provided for them – and from this reconstruct their representational aspects. So by analogy, rather than taking the representational practices (analogues of truth and reference) to explain the inferential capacity of carriers (the analogue of validity), we reconstruct the practices by taking the notion of surrogative reasoning as conceptually basic.

Such an approach is outlined by de Donato Rodríguez and Zamora Bonilla (2009) and seems like a fruitful route for future research. There is no textual evidence that Suárez would endorse such an approach. And, more worrying for Inferentialism 2, it is not clear whether such an approach would satisfy s-deflationism. Each investigation into the inferential rules utilised in each instance, or type of instance of epistemic representation will likely be a substantial (possibly sociological or anthropological) project. Thus the s-deflationary credentials of the approach – at least if they are taken to require that nothing substantial can be said about scientific representation in each instance, as well as in general – are called into question.

Finally, if the conditions in Inferentialism 2 are taken to be abstract platitudes then we arrive at an abstract minimalism, which Suárez associates with Wright. Although Inferentialism 2 defines the concept of epistemic representation, the definition does not suffice to explain the use of any particular instance of epistemic representation because "on the abstract minimalism here considered, to apply this notion to any given concrete case of representation requires that some additional relation obtains between [X] and [T], or a property of [X] or [T], or some other application condition" (Suárez 2015, p. 48; cf. Suárez and Solé 2006). Hence, according to this approach representational force and inferential capacity are taken to be abstract platitudes that suffice to define the concept of scientific, or epistemic, representation. However, because of their level of generality, they fail to explain any particular instance of it. To do this requires reference to additional features that vary from case to case. These other conditions can be "isomorphism or similarity" and they "would need to obtain in each concrete case of representation" (Suárez 2015,

5.3 Motivating Deflationism

p. 45).[10] Suárez calls these extra conditions the *means* of representation, the relations that scientists exploit in order to draw inferences about targets from their models, and they are to be distinguished from conditions (i) and (ii), the *constituents* of representation, that define the concept.[11] We are told that the means cannot be reduced to the constituents but that "all representational means (such as isomorphism and similarity) are concrete instantiations, or realisations, of one of the basic platitudes that constitute representation" (Suárez and Solé 2006, p. 43), and that "there can be no application of representation without the simultaneous instantiation of a particular set of properties of $[X]$ and $[T]$, and their relation" (ibid., p. 44).

Such an approach amounts to using conditions (i) and (ii) to answer the ER-problem, but again with the caveat that they are abstract conditions that require concretisation in each instance of epistemic representation (although in this guise, the concretisation focuses on means rather than practices). In this sense it is immune to Contessa's objection about the "mysterious" capacity that carriers have to licence inferences about their targets. They do so in virtue of more concrete relations that hold between carriers and their targets, albeit relations that vary from case to case. The key question facing this account is to fill in the details about what sort of relations concretise the abstract conditions. But we are now facing a similar problem as the above. Even if s-deflationism applies to epistemic representation in general, an investigation into each specific instance of it will involve uncovering substantial relations that hold between carriers and their targets. Alternatively, one could submit that the abstract conditions should be taken to be explanatorily prior to the more concrete relations. But nevertheless, the project of understanding how these conditions explain the concrete relations in each particular case would amount to a substantial (again possibly sociological or anthropological) project. In both cases, this seems to conflict with Suárez's adherence to an s-deflationist approach.

We conclude that whether deflationism lends support to *Inferentialism 2* is at best an open question. There is, however, another route that inferentialists could pursue in support of their view. The clue comes from Chakravartty's (2010a) classification of accounts of scientific representation. He notes that in the literature on scientific representation there is an important difference between what he calls "informational" and "functional" kinds of views. Views that fall into the former family are taken to characterise a scientific representation as something that "bears an objective relation to the thing it represents, on the basis of which it contains information regarding that aspect of the world" (ibid., p. 198). Views that fall into the latter family "emphasize cognitive activities performed in connection with these targets, such as interpretation and inference" (ibid., p. 197). These views give rise to two different kinds of characterisations one can give of a representation. What we call an *inherent characterisation* gives an account of the relation that a carrier bears to its target; a *functional characterisation* gives an account of the functional roles that representations play in science. The contrast between inherent and functional

---

[10] See also Suárez's (2004, p. 773 and p. 776) and Suárez and Solé's (2006, p. 43).
[11] See his (2003, p. 230; 2010, pp. 93–94; 2015, p. 46) and Suárez and Solé's (2006, p. 43).

characterisations can be illustrated with simple example. Assume you ask the question "what is a boiler?". A functional characterisation describes a boiler as contraption that makes water hot. An inherent characterisation will explain how a boiler achieves this. Such a characterisation will have to distinguish between different kinds of boilers – continuous-flow boilers and storage boilers, gas boilers and electric boilers, combi boilers and systems boilers, and so on – and then give an account of how each of them manages to make water hot.

The justification of *Inferentialism 2* then might be that it offers a functional rather than an inherent characterisation, and that giving either kind of characterisation is equally acceptable – indeed, when presenting an account of epistemic representation, we're free to choose whether we give a functional or an inherent characterisation. So when formulating *Inferentialism 2* we have merely made use of the right to choose. This seems to be in line with Chakravartty's take on his distinction when he emphasises that the two approaches to scientific representation "are in fact complementary" (ibid., p. 199) and "focus on different aspects of one and the same thing: the nature of scientific representation" (ibid., p. 209).[12]

Even if this thesis is correct and the two families of views describe the same thing, it doesn't follow that both are on par, much less that one can freely choose to engage with only one and ignore the other. Heads and tails are, literally, two sides of the same coin, but one can't have one without the other, and a numismatist who decides to look only at one side of the coin only gains partial knowledge and is likely to miss out on crucial information. A functional characterisation is useful when it comes to understanding how an item integrates with other items and contributes to a larger system. When designing a block of flats the architect may well not need to know much more about a boiler other than that it warms water, but the same characterisation is plainly insufficient for an engineer who has to build a boiler. Likewise, to know that an epistemic representation allows competent agents to draw inferences about the target system may well be all that a metaphysician needs to know in order to study what science says about, say, the metaphysics of space and time, but it will not be enough for a physicist who has to build, study, and test a model, or for a philosopher of science who wants to understand how the physicist does these things.

The question then is whether a theory of epistemic representation should serve architects or engineers, as it were. Should it focus on functional roles, or should it focus on how those roles are realised in each particular case? Phrased in this way, this is by and large an open question in the literature. Philosophical tastes vary. Those who prefer to systematise may stay at an abstract level and focus on function while those who prefer to investigate how things work may provide detailed case studies of particular models, but there is little self-reflective justification for which is the best way to proceed. The advantage of the former approach is that it provides us with an understanding of how different models from different sciences, and indeed different carriers from different areas of life, fit together. But this comes at

---

[12] Shech calls this the complementarity thesis (2016, p. 315).

the cost of failing to fully explicate any particular instance epistemic representation. The advantage of the latter is that it provides us with a detailed understanding of how a particular instance of epistemic representation works, but such an understanding remains divorced from how that representation should be understood from a broader perspective concerning how humans come to understand the world via their use of epistemic representations in general.

Our position is that a comprehensive theory of epistemic representation should be able to perform both roles. If the block of flats is to be built in such a way that all of the inhabitants have hot water, then we need to draw upon the architect's functional characterisation *and* the engineer's inherent one. Thus, understanding the functional roles representations play is important, but this cannot *replace* a study of how these roles are realised in concrete cases. By themselves, functional definitions are not enough to gain an understanding of a subject matter that is sufficient to work with, and inferentialism can't appeal to the dichotomy between inherent and functional characterisations to justify restricting focus solely on the two abstract functional conditions in Inferentialism 2.

5.4 Reinflating Inferentialism: Interpretation

In response to difficulties of the kind discussed in the last section, Contessa claims that "it is not clear why we should adopt a deflationary attitude *from the start*" (2007, p. 50, original emphasis) and provides an "interpretational account" of epistemic representation that is inferentialist but without being deflationary. Contessa claims:

> The main difference between the interpretational conception [...] and Suárez's inferential conception is that the interpretational account is a substantial account — interpretation is not just a "symptom" of representation; it is what makes something an epistemic representation of a something else. (ibid., p. 48)

To explain in virtue of what the inferences can be drawn, Contessa introduces the notion of the *interpretation*[13] of a carrier in terms of its target system as a necessary and sufficient condition on epistemic representation (Contessa 2007, p. 57; see also Contessa 2011, pp. 126–27):

Interpretation: a carrier X is an epistemic representation of a certain target T iff there is a user of X who adopts an interpretation of X in terms of T.

Contessa offers a detailed formal characterisation of an interpretation (2007, pp. 57–62). The leading idea of his account is that the user first identifies a set of relevant objects in the carrier X along with a set of properties and relations that these objects instantiate. She then does the same in the target T, where she also identifies

---

[13] "Interpretation" is being used here in a different way both to how it was used in Chap. 4 to discuss linguistic aspects of structures and how it will be used in Chaps. 8-9.

a set of relevant objects along with a set of properties and relations these objects instantiate. The user then (a) takes X to denote T; (b) takes every identified object in the carrier to denote exactly one object in the target (and every relevant object in the target to be so denoted, which results in there being a one-to-one correspondence between relevant objects in the carrier and relevant objects in the target); and (c) takes every identified property and relation in the carrier to denote a property or relation of the same arity in the target (and, again, and every property and relation in the target to be so denoted, which results in there being a one-to-one correspondence between relevant properties and relations in the carrier and target). A formal rendering of these conditions is what Contessa calls an *analytic interpretation*.[14] The relationship between interpretations and the surrogative reasoning mentioned above is that it is *in virtue* of the user adopting an analytic interpretation that a carrier licences inferences about its target.

At first sight Contessa's interpretation may appear to be equivalent to setting up an isomorphism between carrier and target. This impression is correct in as far as an interpretation requires that there be a one-to-one correspondence between relevant elements and relations in the carrier and the target. However, unlike the isomorphism view, Contessa's Interpretation is not committed to carriers being structures, and relations can be full-fledged relations rather than purely extensionally specified sets of n-tuples.

Interpretation is a non-deflationary account of epistemic representation: most (if not all) instances of epistemic representation involve the user of a carrier adopting an analytic interpretation of the carrier in terms of a target, and doing so makes the account substantial.[15] The capacity for surrogative reasoning is then seen as a symptom of the more fundamental notion of the user of a carrier adopting an interpretation of a carrier in terms of its target.

Let's now turn to how Interpretation fares with respect to our questions for an account of epistemic representation as set out in Chap. 1. Interpretation provides necessary and sufficient conditions[16] on epistemic representation (recall the discussion in Sect. 5.1 regarding the demarcation problems) and hence answers the ER-Problem. Furthermore, it does so in a way that explains the directionality of representation: interpreting a carrier in terms of a target does not entail interpreting a target in terms of a carrier.

Contessa does not comment on the Applicability of Mathematics Condition, but since his account shares with the structuralist account an emphasis on relations and one-to-one carrier–target correspondence, Contessa can appeal to the same account

---

[14] Contessa's definition includes an additional condition pertaining to functions in the carrier and target, which we suppress for brevity.

[15] Contessa focuses on the sufficiency of analytic interpretations rather than their necessity, and notes that he does "not mean to imply that all interpretation of vehicles [carriers] in terms of the target are necessarily analytic. Epistemic representations whose standard interpretations are not analytic are at least conceivable" (2007, p. 58). Even with this in mind, it is clear that he intends there being *some* interpretation to be a necessary condition on epistemic representation.

[16] Modulo the qualification mentioned in footnote 15.

5.4 Reinflating Inferentialism: Interpretation

of the applicability of mathematics as structuralist. It remains unclear how Interpretation addresses the Problem of Style. One option available to Contessa would be to classify all instances of analytic interpretation as the same style, and then embark on a sustained investigation into non-analytic interpretations. Another option would be to classify styles according to the types of objects, properties, and relations specified in the analytic interpretations. Both routes require further investigation.

With respect to the Problem of Carriers, Interpretation itself places few constraints on what carriers are, ontologically speaking, and on how they are handled. All it requires is that they consist of objects, properties, and relations. For this reason, the problems discussed in Sect. 3.4 rear their heads again here. As before, how to apply Interpretation to physical carriers can be understood relatively easily. But how to apply it to non-material carriers is less straightforward. Contessa (2010) distinguishes between mathematical models and fictional models, where fictional models are taken to be fictional objects. We discuss fictional objects in Chap. 6.

How Interpretation allows for the possibility of misrepresentation is a subtle issue. Before turning to it directly it is useful to quickly clarify how Interpretation accounts for what looks like misrepresentation but actually isn't. These cases involve carriers that contain aspects that don't, and are known not to, correspond to any aspect of their targets. To deal with these sorts of cases Contessa notes that "a user does not need to believe that every object in the model denotes some object in the system in order to interpret the model in terms of the system" (2007, p. 59). He illustrates this claim with an example of contemporary scientists using the Aristotelian model of the cosmos to represent the universe, pointing out that "in order to interpret the model in terms of the universe, we do not need to assume that the sphere of fixed stars itself [...] denotes anything in the universe" (ibid.). From this example it is clear that the relevant sets of objects, properties and relations isolated by the analytic interpretation do not need to exhaust the objects, properties, and relations, of either the carrier or the target. The carrier user can identify a relevant proper subset in each instance. This allows Interpretation to capture the common practice of abstraction in epistemic representation: a carrier need only represent some features of its target, and moreover, the carrier may have "surplus" features that are not taken to represent anything in the target (that is, features that don't play a direct representational role).

But these cases are not really misrepresentations, because a user who adopts an interpretation of the Aristotelian model according to which the sphere of fixed stars doesn't denote anything in the target doesn't generate any inferences with false conclusions (or at least doesn't generate any such inferences on the basis of the sphere in the model). As in the case of Inferentialism 1, this brings us back to the distinction between sound and valid inferences. If we describe an inference from carrier to target as a "surrogative inference" then "a surrogative inference is sound if it is valid and its conclusion is true of the target. However, a surrogative inference can be valid even if it is not sound (i.e., an inference is valid irrespectively of the truth of its conclusion)" (ibid. p. 51). This allows Interpretation to meet the

Misrepresentation Condition and provide standards of accuracy: an epistemic representation is accurate to the extent that it delivers sound surrogative inferences.

The question then, is what makes a carrier-to-target inference valid according to Interpretation. Here Contessa provides rules linking analytic interpretations with valid surrogative inferences. The rules seem to be motivated by the semantics of predicate logic, which first assigns the symbols of a formal language corresponding objects and properties in a structure, and then determines the meaning of sentences on the basis of these correspondences (recall the discussion of the semantic aspects of structures in Sect. 4.2). Contessa's first rule is that, if, according to the interpretation, an object in the carrier o_c denotes an object in the target o_t, then it is valid to infer that o_t is in the target iff o_c is in the carrier. The second rule is that if, according to the interpretation, an object in the carrier o_c denotes an object in the target o_t, and the property in the carrier P_c denotes the property in the target P_t, then it is valid to infer that o_t has P_t iff o_c has P_c (ibid. p. 61).[17]

We can now distinguish between two different ways in which a carrier-to-target inference can have a false conclusion. First, we can consider cases where the conclusion states that there is an object o_t in the target that is (or isn't) in the extension of a property in the target P_t, but that in the target itself o_t actually isn't (or is) in the extension of P_t (where P_t is instantiated somewhere in the target). For example, the target system might consist of three separate objects, two of which have a property and the third of which doesn't, and the conclusion of the carrier-to-target inference might state that all three objects in the target have that property. In order to accommodate this as an instance of misrepresentation according to Interpretation it must be the case that there is a possible carrier-to-target inference that is valid according to the above rule, but which nevertheless generates the false conclusion (thereby being valid but unsound). The above rules easily accommodate such cases. What is required is an analytic interpretation according to which there is an object in the carrier o_c that denotes the object in the target o_t, a property in the carrier P_c that denotes the property in the target P_t, and that the carrier and the target are mismatched in the sense that either o_t is in the extension of P_t but o_c is not in the extension of P_c, or vice versa. For example, the carrier might consist of three objects, all sharing a property P_c, and the target consist of three objects, of which only two have the property P_t. Then assuming an interpretation according to which each carrier object denotes a distinct target object, and that P_c denotes P_t, the above rules will generate the valid inference that all of the target objects are in the extension of P_t, which will be the desired valid-but-unsound inference required in order for Interpretation to handle these kinds of instances of misrepresentation.

However, as Shech (2015) points out, there are other kinds of misrepresentation that Interpretation and the above rules have difficulty dealing with. In particular, these are cases where the misrepresentation concerns using a carrier to generate a claim according to which there are objects in the target that have a property P_t even though P_t isn't instantiated in the target at all (notice the difference with the above

---

[17] Contessa's rules also cover relations and functions, which we suppress here.

5.4 Reinflating Inferentialism: Interpretation

kind of case, where the property was instantiated somewhere in the target but so that there was just a carrier–target mismatch). His example is using an equilateral triangle (a triangle with three sides of equal length and three equal interior angles) to (mis)represent an obtuse triangle (a triangle with one angle that is greater than 90 degrees) in the sense of generating the false claim that the obtuse angles in the target triangle are all equal to one another. He argues that a valid carrier-to-target inference to such a conclusion cannot be generated by Interpretation. The reasoning is straightforward: in order to generate the false claim that some object, o_t, is in the extension of some property, P_t, in the target, where the claim is false because P_t doesn't exist in the target at all, it must be the case that some property in the carrier denotes P_t. But if P_t doesn't exist in the target, then it cannot be denoted by any property of the carrier. So, such a valid but unsound inference cannot be generated, and thus Interpretation cannot handle cases of misrepresentation where some property is claimed to be present in the target, which in fact isn't.

This problem finds an elegant solution in Díez's (2020) "Ensemble-Plus-Standing-For" account of representation. Díez considers carriers to be an ensemble, where an ensemble is a structured collection of objects with properties and relations (including higher-order ones) which have "appropriate logical form" (ibid. p. 139). This condition rules out mismatches concerning the number of objects in the collection and the arity of the relations in question, as well as the order of the relations in question. For instance, two individuals standing in a first-order binary relation; two individuals each having a first-order monadic property; or two first-order monadic properties standing in a second-order binary relation are ensembles. However, two individuals and a ternary relation are not an ensemble because there is no way in which two individuals can stand in ternary relation (at least when the ternary relation is of the sort that only holds between distinct individuals). Similarly, two objects and a second-order binary relation cannot constitute an ensemble because the objects are not appropriate for the order of the relation in question.

There is no presupposition that an ensemble is abstract; both physical objects and abstract entities can constitute ensembles in this sense. A model then is an ensemble plus a standing for relation. The standing for relation consists in a model user taking the constituents of the model to stand for constituents of the target so that "logical congruency" is preserved (ibid. p. 140).[18] The condition of logical congruency essentially means that the constituents of the target domain can be arranged to form an ensemble of the same type as the model. The easiest way to achieve this is to constrain the standing for relation so that entities in the model stand for entities of the "same logical category" in the target: individuals stand for individuals, first-order dyadic relations stand for first-order dyadic relations, and so on (although

---

[18] In this sense Díez denies that anything can represent anything else, as discussed in Sect. 2.3. On his account, a "simple" (e.g. a salt shaker taken as simple) cannot represent, not even wrongly, something complex (e.g. Madagascar taken as complex), and an individual instantiating a monadic property cannot be used to represent a target that is taken to have the logical form of two individuals in a binary relation.

there are other ways of achieving this, as explained shortly). Thus understood, Díez's notion of standing for, constrained by intended logical congruency, is extremely close to Contessa's Interpretation (where the requirement of congruency was left implicit), and the two accounts share the notion that the core of representation is to interpret one constituent in terms of another constituent subject to certain constraints. But unlike Contessa, Díez does not restrict modelling to first-order relations. He allows for ensembles to have higher-order features, and for logical congruency to cover cases where the denoted constituents are not of the same logical category as the corresponding model constituents. For instance, logical congruence can be achieved if two individuals instantiate a dyadic first-order relation in the model and if the first-order relation stands for a second-order dyadic relation and the two individuals stand for two monadic first-order properties.

The crucial difference between Díez's and Contessa's accounts is that Díez distinguishes between "performance conditions" and "adequacy conditions" for representation (which he also refers to as "existence conditions" and "success conditions"). Díez develops this idea in analogy with theories of speech acts. A speech act is performed only if it meets certain conditions. Yet, and this is the crucial point, even if the act is performed correctly, it can still fail in the sense of not achieving its goals. Hence, in addition to meeting performance conditions, a speech act also must satisfy success conditions to have its effect (ibid., p. 136). Díez submits that this distinction is equally relevant in the case of epistemic representation. The performance conditions for an agent to use X to represent T are that (1) both X and T are ensembles; (2) that the agent intends to use model X to represent target T; (3) that the agent takes the (contextually relevant) constituents of X to stand for constituents of T so that logical congruency is preserved; and (4) that all (contextually relevant) constituents of T are stood for by a (contextually relevant) constituent of X (ibid., p. 140). The intuitive idea behind these conditions is that for an epistemic representation to be established, the agent must conceptualise a target phenomenon as an ensemble and build a model ensemble (in which she attributes logical categories to the constituents of the model and the target) in such a way that the epistemic representation is: logically coherent (it satisfies the logical congruency constraint); materially coherent (taking the stood for objects to belong to the previously selected target); and complete (does not leave out any relevant constituents of the target). These performance conditions are Díez's answer to the ER-Problem. The success conditions add to the performance conditions that (i) the postulated intended target entities exist; (ii) that the stood for entities in T behave as their corresponding entities in M; and (iii) that the agent's purposes are served by using X to represent T in virtue of (ii) (ibid., p. 142).

It is clear that the performance conditions can be satisfied while some, or even all, success conditions fail. If this happens, then X is a misrepresentation. And, coming back to our discussion of Shech's objection to Contessa, Díez's account can handle cases of the kind Shech describes: an agent can *intend* to represent a feature in T by matching it up with a feature of adequate logical category in X (and hence

5.4 Reinflating Inferentialism: Interpretation

satisfy the performance condition) but fail to do so because that feature turns out not to exist (and to fail the first success condition).[19]

Taking this line of argument introduces another possible problem, highlighted by Bolinska (2013). She argues that there are cases where a model user can offer an analytic interpretation of the carrier (or performance conditions of the kind Díez envisages) in terms of the target that the model user clearly doesn't believe will generate true conclusions. For example: a model user could use the Rutherford model of the atom to generate conclusions about a hockey puck on the ice by taking the electron in the model to denote the puck and the nucleus the surface of the ice, and from this infer that the puck is negatively charged, which is clearly false. This would be a case where the inference in question is valid and not sound, thereby making the model a misrepresentation. Now, according to Bolinska, in such a case

> the question is not whether the model is a *faithful* epistemic representation of its target, but whether it is an epistemic representation of its target *at all*. It may well be the case that no valid inferences from the model to the target system are sound, but even in this case [according to Interpretation], the model is still an epistemic representation. (ibid., p. 227, origional emphasis)

She concludes that such a case should not be deemed an epistemic representation, and in doing so argues that Interpretation is not a tenable response to the ER-Problem. This, if correct, equally calls into question Díez's performance conditions. She then goes onto argue that epistemic representations require that the user choose (and interpret) a carrier "with the aim of faithfully representing (this aspect of) the target" (ibid., p. 228). If one were to build this into Contessa's and Díez's accounts, then this would amount to the model user at least believing that the hypothesized carrier–target relation generates sound inferences according to the aforementioned rules.

It is not obvious to us that the aim of faithfully representing the target should be built into the definition of epistemic representation. In the case in question, why shouldn't we accept that the Rutherford model is an epistemic representation of the hockey puck? It does allow for surrogative reasoning about it, and in general it is not obvious that epistemic representation by a user commits that user to believing that the representation is accurate (cf. the discussion in Sect. 4.5 and Nguyen 2016). Moreover, there are many cases of epistemic representation in science where the model user knows their model is inaccurate at least in certain respects. Contessa illustrates this with the example of a massless string. The model user knows that the

---

[19] Díez's account could be re-described as an account in which the model user formulates a *hypothesis* (in the sense discussed in Sect. 3.2) that the model and the target stand in the right relation to one another, a hypothesis which can then turn out to be either true or false. Shech (2015) does consider such a response but rejects it for reasons we find unconvincing. He argues that if such a move were allowed, then one could use a carrier to misrepresent itself (by offering an interpretation hypothesizing that it has properties it doesn't). He claims that such a scenario allows for valid but unsound inferences, but that this doesn't match up with misrepresentation, since nothing can misrepresent itself. This doesn't seem true: we can consider an altered Magritte painting with the inscription: "ceci est une pipe". This is clearly a case of self-misrepresentation, and Shech offers no arguments for why self-misrepresentation is impossible in the contexts he's considering.

model is inaccurate, but appeals to a user's corrective abilities to justify their use of it: "Since models often misrepresent some aspect of the system or other, it is usually up to the user's competence, judgment, and background knowledge to use the model successfully in spite of the fact that the model misrepresents certain aspects of the system" (Contessa 2007, p. 60).

Ducheyne (2012) provides a variant of Interpretation that has a similar motivation to Bolinksa's. The notion of an analytic interpretation allows for arbitrary connections between aspects of carriers and aspects of their targets (Bolinska's specific objection is that it is these arbitrary connections that allow for interpretations of carriers in terms of targets that she takes to be not epistemic representations at all, rather than drastic misrepresentations).[20] Ducheyne therefore tries to constrain the sorts of connections between carrier and target features that analytic interpretations, at least in science, should make. The central idea is that it develops Interpretation with the requirement that each relevant relation specified in the interpretation, holds precisely in the model – a condition that Ducheyne explicates by saying that the model is "an ideal *ceteris paribus* and *ceteris absentibus* state of affairs" (ibid., p. 83) – and corresponds to the *same* relation in the world which, however, holds only approximately (with respect to a given purpose) in the target.[21] For example, the low mass of an actual pendulum's string approximates the masslessness of the string in the model. The one-to-one correspondence between (relevant) objects and relations in the model and target is retained, but the notion of a user taking relations in the model to denote relations in the target, is replaced with the idea that the relations in the target are approximations of the ones they correspond to. Ducheyne calls this the *Pragmatic Limiting Case* account of scientific representation (the pragmatic element comes from the fact that the level of approximation required is determined by the purpose of the model user). This imposes stringent constraints on possible interpretations and thereby undercuts cases like Bolinska's hockey puck.

However, if this account is to succeed in explaining how distortive idealisations are scientific representations, then more needs to be said about how a target relation can "approximate" a model relation. Ducheyne implicitly relies on the fact that relations are such that "we can determine *the extent to which* [they hold] empirically" (2012, p. 83, emphasis added). This suggests that he has quantifiable relations in mind, and that what it means for a relation r in the target to approximate a relation r' in the model is a matter of comparing numerical values, where a model user's purpose determines how close they must be if the former is to count as an

---

[20] Ducheyne (2012, p. 80) also claims that Contessa's account fails to distinguish scientific representation from epistemic representation more generally (and he offers an example of using colours in a picture of a group of runners to represent the finishing position of each runner). According to Ducheyne, the example "meets Contessa's requirements for analytic interpretation, but it does not seem to qualify as a scientific representation" (ibid.). Since, as noted in Sect. 5.1, Contessa is not attempting to demarcate scientific representation from epistemic representation more generally this is not a criticism of Contessa's proposal.

[21] Following Ducheyne we use "model" rather than "carrier" to indicate that his discussion is focused on scientific models, rather than epistemic representations more generally.

approximation of the latter. But whether this exhausts the ways in which relations can be approximations remains unclear. Hendry (1998), Laymon (1990), Liu (1999), Ramsey (2006), and Norton (2012) among others, offer discussions of different kinds of idealisations and approximations, and we discuss a specific kind of limit idealisation in Sect. 9.3. Ducheyne would have to make it plausible that all these can be accommodated in his account.

More importantly, Ducheyne's account has problems dealing with misrepresentations. Although it is designed to capture models that misrepresent by being approximations of their targets, it remains unclear how it deals with models that are outright mistaken. For example, it seems a stretch to say that Thomson's model of the atom is an approximation of the structure of atoms, and it seems unlikely that there is a useful sense in which the relations that hold between electrons in Thomson's model "approximate" those that hold in reality. But this does not mean that Thomson's model is not a scientific, or epistemic, representation of the atom; it's just an incorrect one. It does not seem to be the case that all instances of scientific misrepresentation are instances where the model is an approximation of the target (or even conversely, it is not clear whether all instances of approximation need to be considered cases of "misrepresentation" in the sense that they licence falsehoods about their targets).

Finally, the account seems to overshoot its goal and rule out perfectly respectable scientific representations just because they involve conventional associations. Consider the following examples (Frigg and Nguyen 2017c, p. 52) which we also discuss in more detail in Chap. 8. After a short immersion in a solution, a strip of litmus paper exemplifies a certain shade of red, which can easily be converted into the solution's pH value using the paper's translation manual. The particular shade of red represents a particular level of acidity. But shades of red are not approximations of acidity by any standard of approximation. Or consider a standard representation of the Mandelbrot set (see Fig. 8.5 in Chap. 8). The square shown is a segment of the complex plane, and each point represents a complex number c. This number is used as parameter value for an iterative function. If the function converges for number c, then the point in the plane representing c is coloured black. If the function diverges, then a shading from yellow to blue is used to indicate the speed of divergence, where yellow is slow and blue is fast. This colour scheme offers an epistemic representation that contains much valuable information, but without colours being approximations of convergence speed.

Chapter 6
The Fiction View of Models

Scientific modelling involves a creative act of the imagination. We have seen in the Introduction that Newton, when modelling the solar system, considered a system consisting of perfect spheres with homogenous mass distributions gravitationally interacting with each other in otherwise empty space. The fact that modelling involves an act of the imagination, and that at least parts of the content of what is imagined deviates from reality, motivates a family of approaches that analyse modelling and representation by drawing analogies with literary fiction.

We begin by outlining the guiding intuitions behind the fiction view and by pinpointing the problems that it raises (Sect. 6.1). Doing so reveals that there is no such thing as "the" fiction view. In fact, there are two branches of the view that construe the analogy between models and fiction differently. Positions in the first branch construe the analogy narrowly and see the fiction view as providing a response to the Problem of Carriers while leaving all other problems concerning scientific representation untouched. Positions in the second branch construe the analogy more broadly and see the fiction view as providing a comprehensive theory of representation. Either construal has to get clear on what fiction is before it can develop the analogy. We distinguish two senses of fiction: fiction as infidelity and fiction as imagination and note the latter is the relevant sense of fiction in the current context (Sect. 6.2). The next four sections discuss the first branch. To pave the ground for a discussion of the Problem of Carriers, we begin by detailing the questions that an account of carriers has to answer in the context of the fiction view of models (Sect. 6.3). We then introduce pretence theory (Sect. 6.4). With this in place, we formulate an explicit account of carriers as fictions (Sect. 6.5), and say what challenges it faces (Sect. 6.6). We then turn to the second branch and spell out how the analogy can be put to use to provide a comprehensive account of epistemic representation (Sect. 6.7). To conclude we discuss fundamental criticisms of fictional approaches to modelling, namely ones that take issues with the basic intuitions behind the view and think that likening scientific models to fictions in any way is a mistake (Sect. 6.8).

6.1 Intuitions and Questions

As we have seen in Chap. 1, some scientific models are material objects. Scale models of ships, ball-and-stick models of molecules, and water pipe models of an economy belongs this group of models. But many, if not most, models are not of this kind. They are something scientists hold in their heads rather than in their hands. Accordingly, scientific discourse is rife with passages that *appear* to be descriptions of systems in a particular discipline, and yet there does not seem to be anything in the world that corresponds to these descriptions. Students of mechanics learn about oscillations by investigating the dynamical properties of the so-called ideal pendulum, a pendulum that consists of an extensionless pendulum bob that is suspended from a pivot on a massless and perfectly linear string. Population biologists study the evolution of one species that reproduces at a constant rate in an unchanging and unlimited environment. And when studying the exchange of goods, economists consider a situation in which there are only two goods, two rational agents, no restrictions on available information, no transaction costs, no money, and dealings are done immediately. The surface grammar of such descriptions is the same as the description of ordinary physical systems. Nevertheless, no one would mistake such descriptions for descriptions of *actual* systems: we know very well that there are no such systems. Such descriptions introduce models. Scientists sometimes express this fact by saying that they talk about "model-land" when they talk about things like massless stings.[1]

To come to a better understanding of what it means to hold a model in your head rather than in your hands it is helpful to distinguish between model-descriptions and model-systems. Model-descriptions are the passages in research papers, textbooks, and other scientific publications that describe a model. We here focus on the case in which this description is verbal, but in principle it can take any form (it could also be a schematic drawing or a diagram, for instance). The model-system is the object that the model-description – purportedly – introduces: spherical planets, isolated populations, and so on. The puzzle with non-material models then is that there does not seem to be a system that the model-description describes – that is, there does not seem to be an object of which the model-description is a true description. For this reason, Thomson-Jones (2010, p. 284) refers to model-descriptions as "descriptions of a missing system". Descriptions of this kind are embedded in what he calls the "face value practice" (ibid. p. 285): the practice of talking and thinking about these systems as if they were real. We observe that the amplitude of an ideal pendulum remains constant over time in much the same way in which we say that the moon's mass is approximately 7.34×10^{22} kg. Yet the former statement is about a point mass suspended from a massless string while the latter concerns a real celestial body.

To foreshadow a discussion to come, it's important to note here that analysing the face value practice in terms of "missing" systems does not preclude that at least some parts of a model–description might be true of an actual system in the world

---

[1] See, for instance, Smith's (2007, p. 135).

too, and that even those parts that are not true of an actual system may relate to an actual system in some way or another. Rather, the idea is that, at least at a certain stage of a scientific investigation, the question whether there is such an actual system is simply bracketed. Investigating the implications of the model-description is done in terms of the missing system, independently of whether or not the description, at least in part, is also true of, or related to, some actual system.

Descriptions of missing systems and the face value practice have a familiar ring to them. We not only encounter them in the context of modelling, but also in the context of literary fiction. A novel like Dostoevsky's *Crime and Punishment* describes in detail the mental anguish and moral dilemmas of Rodion Raskolnikov, and the reader is urged to explore the question of how Raskolnikov should be punished for his crime. Dostoevsky's novel is a description of the inner life and actions of a person, and yet the person is not a real person. Raskolnikov is a fictional character. Adapting Thomson-Jones' terminology slightly, one can say that the text of the novel is a "description of a missing person" and that the readers' deliberations about Raskolnikov's guilt are an instance of the face value practice.

We now have in front of us the point of departure of the fiction view of models: model-descriptions are like the texts of literary fiction, and model-systems are like fictional characters.

This parallel between models and fiction has not gone unnoticed. To derive his equations of the electromagnetic field, James Clerk Maxwell first discussed the motion of a "purely imaginary fluid" (Niven 1965, pp. 159–60) and then studied its dynamical properties in detail. To arrive at the equivalence principle of his general theory of relativity, Albert Einstein invites the reader to first "imagine a large portion of empty space" and then "imagine a spacious chest resembling a room with an observer inside" located in this portion of empty space (Einstein 1920/1999, p. 60). Maxwell and Einstein did not take themselves to be describing real physical situations and they highlighted the fictional character of the systems they considered by describing them as the result of an act of the imagination.[2]

Philosophers have followed suit. The parallel between science and fiction occupied centre stage in Vaihinger's (1911/1924) philosophy of the "as if", and Fine (1993, p. 16) notes that modelling natural phenomena in every area of science involves fictions in Vaihinger's sense. Frigg emphasises the involvement of the imagination in modelling and sees models as "imagined objects" (2003, p. 87). The parallel has also been drawn specifically between models and fiction. Cartwright observes that "a model is a work of fiction" (1983, p. 153) and later suggests an analysis of models as fables (1999a, Chap. 2); both Hartmann (1999) and Morgan (2001) emphasise that stories and narratives are integral aspects of models;[3] and a

---

[2] These are not cherry-picked instances, and examples can be multiplied. For further examples see Salis and Frigg's (2020).

[3] Again, these aren't isolated instances. As we have seen in Sect. 3.4, Giere likens models to fiction. McCloskey regards economists as "tellers of stories and makers of poems" (1990, p. 5). Sklar highlights that describing systems "as if" they were systems of some other kind is a royal route to success (2000, p. 71). Morgan (2004) stresses the importance of imagination in model building.

number of authors have recently developed different versions of the fiction view of models (which we discuss later in this chapter).

The parallel between models and fiction has to do real work if we want it to be more than a philosophical conversation piece. But what work would one expect the parallel to do and what philosophical problem is it supposed to solve? Opinions on this are divided, and the fiction view splits into two branches over this question. Authors belonging to the first branch see the analogy with fiction as providing an account of models, the carriers of representations. Models, on that view, are on par with literary fiction as far as an understanding of the carriers of a representation is concerned, while there is no productive parallel between models and fiction as regards the other problems concerning scientific representation. This means that approaches belonging to the first branch only address the Problem of Carriers and see all other problems concerning scientific representation that we have seen in Chap. 1 as being solved independently of the analogy between models and fiction. Getting clear on the nature of the objects that end up "doing the representing" is seen as prolegomena to a theory of representation, but without itself providing, or even constraining, such a theory. For this reason, the discussion in Sects. 6.3, 6.4, 6.5, and 6.6 marks a change in focus from the previous chapters, where the primary focus was on epistemic representation.

Authors belonging to the second branch disagree with the restriction to the Problem of Carriers and see the parallel between models and fiction as providing a response to all problems concerning scientific representation, and in particular to the ER-Problem.[4] According to the approaches in the second branch, understanding models as fictions provides a full-fledged account of scientific, and indeed epistemic, representation. We discuss the first branch in Sects. 6.3, 6.4, 6.5, and 6.6, although Sect. 6.4 introduces a particular approach to fiction that is also important the second branch, namely Walton's pretence theory. The second branch is discussed in Sect. 6.7. But before embarking on a discussion of different uses of fiction, we have to get clear on what is meant by "fiction" and put some basic distinctions in place. This is the task for the next section.

Sugden (2000) points out that economic models describe "counterfactual worlds" constructed by the modeller. Grüne-Yanoff and Schweinzer (2008) emphasise the importance of stories in the application of game theory. Achinstein notes that "[i]n imaginary models an object or system is described by a set of assumptions" and that "[t]he proponent of the model does not commit himself to the truth, or even plausibility of the assumptions he makes; nor does he intend them as approximations to what is actually the case." (1968, p. 220). Davies (2007) draws parallels between issues central to the philosophical literature and questions concerning scientific thought experiments.

[4] Certain uses of fiction in science belong to neither of these branches. Cartwright (2010) uses the literary genres of fables and parables to analyse idealisation, and Sugden (2000, 2009) introduces what he calls credible worlds to understand how theoretical models aid the understanding of real-world phenomena. For a discussion of these approaches see our (2017a, Sect. 3.6.3).

6.2 Two Notions of Fiction

The claim that a model is like a fiction could raise eyebrows because it could be understood as branding models as untrue fabrications, and the sciences in which they occur as wilful concoctions. This flies in the face of an understanding of science as an endeavour aimed at discovering how things are, and it would seem to play into the hands of irrationalists and science-sceptics. However, first appearances notwithstanding, the fiction view of models is not committed any kind of postmodern nihilism. To see why this is we have to clarify the notion of fiction that is at work in the fiction view of models.

Two usages of the term "fiction" must be distinguished: fiction as infidelity and fiction as imagination.[5] Let us discuss each of these notions in turn. In the first usage, something is qualified as a "fiction" if it deviates from reality. The nature of this deviation depends on what is qualified as fiction. If we qualify a *sentence* (or proposition) as a fiction, the relevant kind of deviation is falsity: the sentence is a fiction if it is false. We appeal to this sense of fiction if we say "the prime minister's account of events is complete fiction" to express the view that the prime minister's report of what happened is untrue. If we qualify an *object* as a fiction, the relevant kind of deviation is non-existence: the object is a fiction if it does not exist. We use "fiction" in this sense if we say "the Iraqi weapons of mass destruction were a fiction" to express that there are no, and never were, such weapons.

In as far as objects are qualified as fictions, the sense of existence involved is physical existence in space and time.[6] Hamlet and Emma Bovary have no physical existence, which is why they are *fictional characters*. Yet, there is a pervasive intuition that they *somehow* exist: we think about them, make claims about them, discuss their actions, and so on, which would be not be possible if they were simply nothing. But how should we characterise the "mode of being" of Hamlet and Emma Bovary, and how is discourse about them to be understood? This is a vexing question and we return to it later in Sects. 6.3 and 6.4.

In the second usage, "fiction" applies to a text and qualifies it as belonging to a particular genre, literary fiction, which is concerned with the narration of events and the portraiture of characters. Novels, stories, and plays are fictions in this sense.[7]

---

[5] The notion of fiction as infidelity can found in most dictionaries. The online version of Oxford Living Dictionaries, for instance, defines a fiction as "something that is invented or untrue". The idea that fiction is defined in terms imagination is developed in Evans' (1982) and Walton's (1990). Currie (1990) defines fiction as the result of communicative act on the part of the author, which, like imagination, does not presuppose falsity. For a general discussion of fiction and falsity see Lamarque and Olsen's (1994), and for a specific discussion in the context of scientific modelling Frigg's (2010c, Sec. 1).

[6] We use "exist" in a timeless sense: Beethoven exists, the Second Spanish Republic exists, and the Big Crunch (or the Big Freeze, or the Big Rip, or whatever happens at the end of the universe) exists.

[7] We here focus on written text, but this notion of fiction can be extended to graphic novels, stage performances, radio plays, movies, and different kinds of visual art.

Rife prejudice notwithstanding, falsity is not the defining feature of literary fiction. Neither is it the case that everything that is said in a novel untrue: historical novels, for instance, usually contain correct factual information. Nor does every text containing false reports qualify as fiction: a wrong news report or a flawed documentary do not become fictions on account of their falsity – they remain what they are, namely wrong factual statements. What makes a text fictional is not its falsity (or a particular ratio of false to true claims), but the attitude that the reader is expected to adopt towards it. There is a question about how exactly this attitude should be characterised, but in essence it is one of imaginative engagement.[8] Readers of a novel are invited to imagine the events and characters described. They are expressly not meant to take the sentences they read as reports of fact, let alone as false reports of fact. Someone who reads Tolstoy's *War and Peace* as report of fact and then accuses Tolstoy of misleading readers because there was no Pierre Bezukhov simply doesn't understand what a novel is. This is of course not to say that fiction in this sense is incompatible with truth. The prescription to imagine something is completely compatible with the content of the imagination being true, and in fact even with the reader knowing the content to be true. *War and Peace* contains plenty of correct historical information, and readers are aware of this. Yet, the purpose of the novel is to get readers to imagine specific things, which is independent of (and indeed indifferent towards) the truth and falsity of the content of the imagination.

The two usages of "fiction" are neither incompatible nor mutually exclusive. In fact, many of the places and persons that appear in literary fiction are also fictions in the first sense because they do not exist. But compatibility is not identity. From the fact that something appears in a fiction of the second kind one cannot – and must not – automatically infer that it also is a fiction in the first sense. Pierre Bezukhov and Napoleon both appear in *War and Peace*, but only the former is also a fiction the first sense; it would be a grave error to infer that Napoleon does not exist because he appears in a work of fiction.

Returning to the fiction view of models, the crucial point to note is that the notion of fiction involved in *both* branches of the view is the second notion. When models are likened to fictions, this is taken to involve the claim that models prescribe certain things to be imagined while remaining noncommittal about whether or not the entities or processes in the model are also fictions in the first sense. Just as a novel can contain characters that exist (spatiotemporally) and ones that don't, and prescribe imaginings that are true and ones that are false, models, understood as fictions, can feature existing and non-existing entities alike, and they can ground true and false claims. This is why the fiction view is not guilty of any kind of nihilism. An assessment of which of a model's elements exist and of which of its claims are true is in no way prejudged by the fact that a model is a fiction. Different versions of the fiction view do this in different ways, and we will discuss these ways later in this

---

[8] Imagination can be propositional and need not amount to producing mental pictures. For a discussion of the notion of imagination with a special focus on imagination in scientific modelling see Murphy's (2020), Salis and Frigg's (2020), and Stuart's (2017, 2020). For a discussion of how scientists can imagine allegedly unimaginable models see McLoone's (2019).

chapter. But all versions of the fiction view of models share a commitment to the second sense of fiction and do not mean to brand models as falsities when they say that they are fictions.

One might worry that this is in tension with the face value practice introduced earlier. If the motivation for the fiction view of models is that we talk about a model as though it were real even though we know that there is no actual system of which the model-description is true, then despite invoking fiction as imagination rather than fiction as infidelity, isn't it the case that infidelity is required to account for the face value practice? Not quite. As mentioned previously, thinking about model-descriptions as describing missing systems doesn't preclude them from, at least in part, being true of actual systems. The motivation is rather that, at least in the context of model building and investigation, the question of whether model-descriptions additionally describe, or are related in some way to, an actual system is bracketed. The reason for doing so is to gain creative freedom. Scientists need to have the freedom to play with assumptions, to consider different options, and to ponder variations of hypotheses. In doing so they bracket the question whether the model-objects correspond to real-world objects for a moment and investigate a certain scenario which seems interesting enough to have a closer look. But this does *not* commit them to believing that the scenario is completely false when taken to be about a real-world target. In fact, what motivates scientists to consider a certain scenario often is that they think that the scenario has *something* to do with the target system that they are interested in. However, what that something is will be assessed only ex post facto, once all the details in the model are worked out. Fibonacci didn't consider a population of immortal rabbits thinking "oh, it's all false but it's fun". Fibonacci's motivation to consider such a population was that he would eventually be able to learn something from the model that is true of real rabbits. But to be able to say what the true bits are, he first had to work out what happens in the model. The fiction view of models utilises fiction as imagination to understand this practice, and this remains compatible with the idea that model-descriptions are, at least in part, true (i.e. not fictional in the sense of infidelity) when applied to an actual system.[9]

The fact that the relevant notion of fiction for models is fiction as imagination does of course not mean that the notion of fiction as infidelity plays no role in science. In fact, there are cases where products of science are qualified as fictions in just that sense. To discuss such cases in a systematic way, it is helpful to distinguish two cases. In the first case an infidelity-fiction is a counterfeit, forgery, or fake, produced with the deliberate intention to deceive and mislead. A conman makes people believe that he's from an influential and wealthy family to gain their trust. Yet his pedigree is a fabrication with the sole purpose to persuade people to do things for him that they would not otherwise do. In the second case we entertain a supposition that is known by everybody involved to be at variance with fact, and we do so because it serves a certain purpose. We know that Santa Claus does not exist, yet we

---

[9] Usually at best parts of such descriptions are true of actual systems, but nothing rules out the existence of limiting cases where even the entire description can be true. We will say more about this below.

act as if Santa came to town and organise celebrations because accepting the Santa Claus fiction serves social functions: it creates an opportunity to share gifts, gather the family for a celebration, and so on.

The first case of infidelity-fiction, deliberate deceit, plays no intrinsic role in science and only occurs when there is foul play. In an ideal world there would be no such fictions. But we don't live in an ideal world, and scientists don't always behave as they should. Biologist Hwang Woo-Suk reported to have created human embryonic stem cells by cloning. However, it turned out that he deliberately lied to the public and that he managed to keep up appearances with an elaborate system of secrecy and fraud: his human embryonic stem cells were fictions. After the charade unravelled, he was put on trial and sentenced to two years in prison.

Something similar happened in the Theranos scandal, now widely thought to be one of Silicone Valley's greatest disasters. Theranos Founder and CEO Elizabeth Holmes made investors believe that her company was in possession of life changing new blood testing technology. The "nanotainer" was a new blood collection vessel that collected small amounts of blood from a little puncture in the finger. This small blood sample was then said be analysed by a machine no larger than a desktop printer, the "Edison", which, assisted by remote access technology, was supposed to run hundreds of diagnostic tests from just these few droplets of blood. Theranos was able to raise more than 700 million dollars of venture capital and in 2014, and at the peak of its popularity, was valued at nearly 10 billion dollars. The tides changed in October 2015 when an article in the *Wall Street Journal* questioned the validity of the technology and investigators found out that the technology didn't exist and that Theranos carried out its blood tests behind the scenes on conventional equipment. The Edison, or at least the Edison as a fully functioning blood analysis machine, existed only by name – it was a fiction (or the first kind)! At the time of writing Holmes and her partner Ramesh Balwani were awaiting trial for fraud and faced a sentence of up to 20 years in prison.

The second case of infidelity-fiction plays a productive role in science, and it doesn't get its progenitors into jail. Suppositions known to be at variance with fact are often accepted because they serve a certain purpose (and one of them is to play a role in scientists' imagination). D'Alembert's Principle in classical mechanics furnishes an example (Kuypers 1992, pp. 13–22). The task at hand is to predict the trajectory of a particle whose motion is limited by external constraints that vary over time. This happens, for instance, when we try to calculate the motion of a marble in a moving bowl. To solve this problem D'Alembert introduced the concept of a *virtual displacement*, an infinitesimal but infinitely fast displacement of the particle compatible with the constraints, and he postulated that the nature of the constraints be such that the virtual displacements do no work on the system. It follows that the differences between the forces acting on a system and the time derivatives of the momenta of the system itself along a virtual displacement consistent with the constraints is zero. This posit, now known as D'Alembert's Principle, is a powerful tool to calculate the path of objects moving under time-dependent constraints. But there are no virtual displacements, and, as their name indicates, scientists of course know this – virtual displacements are fictions (of the first kind).

6.2 Two Notions of Fiction

We encounter similar situations in other domains. In thermodynamics transformations are considered to be quasistatic: they are infinitely slow and only pass through equilibrium states. There are no such transformations in a real physical system; quasistatic transformations are expedients to do calculations. In solid state physics the units of vibrational energy that arise from oscillating atoms in a crystal are described as if they were particles, so-called phonons. This facilitates the theoretical treatment of crystals, but there are no particles, which is why they are also referred to as "quasiparticles". In early atomic physics electrons were assumed to move in exact orbits, much like the orbits of a planet moving around the sun. But classical electron orbits have turned out to be fictions. This, however, does not render them useless. In fact, Bokulich (2009) argues that these orbits perform an important explanatory function, and hence are, their fictional character notwithstanding, by no means obsolete.[10]

So far the status of a fiction has been conferred on particular elements of science: virtual displacements, quasistatic transformations, phonons, and classical electron orbits. Depending on where one stands in the realism versus antirealism debate, one may not want to limit fictions in this way and subsume the entire theoretical machinery of science under the notion of fiction (in the sense of benign infidelity). Scientific realists hold that mature scientific theories provide an at least approximately true account of the parts of the world that fall within its scope. Antirealists disagree and submit that we should only take claims about observables at face value and, depending on the kind of antirealism one advocates, either remain agnostic about, or downright renounce commitment to, the theoretical claims of a scientific theory. In the current idiom, the antirealist regards the theoretical posits as fictions. Fine advocates such a position and calls it *fictionalism*:

> "Fictionalism" generally refers to a pragmatic, antirealist position in the debate over scientific realism. The use of a theory or concept can be reliable without the theory being true and without the entities mentioned actually existing. When truth (or existence) is lacking we are dealing with a fiction. Thus fictionalism is a corollary of instrumentalism, the view that what matters about a theory is its reliability in practice, adding to it the claim that science often employs useful fictions. (1998, p. 667)

Being a fictionalist about something then means being an antirealist about a certain domain of discourse, and the domains that can be, and have been, given a fictionalist treatment range from morality to mathematics. A discussion of fictionalism in this broader sense is beyond the scope of this chapter. We just emphasise, again, that it is not fictionalism in this sense that's at work in the fiction view of models. To make this clear, we keep using the moniker "fiction view of models" rather than the more succinct "model fictionalism" or "fictionalism about models".[11]

---

[10] Further examples of fictions of this kind are discussed in Bokulich's (2008) and some of the contributions to Suárez's (2009).

[11] For a general introduction to fictionalism see Salis' (2014). Liu (2016) uses the term "new fictionalism" to refer to the position we call the fiction view of models, and calls fictionalism in Fine's sense "traditional fictionalism". For further discussions of fiction and fictionalism in philosophy of science see Cartwright's (2010), Fine's (1993), Morrison's (2009), Purves' (2013), and Woods'

6.3 Fiction and the Problem of Carriers

The core idea of the first branch of the fiction view of models is clearly expressed in the following passage by Godfrey-Smith:

> [...] I take at face value the fact that modelers often *take* themselves to be describing imaginary biological populations, imaginary neural networks, or imaginary economies. [...] Although these imagined entities are puzzling, I suggest that at least much of the time they might be treated as similar to something that we are all familiar with, the imagined objects of literary fiction. Here I have in mind entities like Sherlock Holmes' London, and Tolkien's Middle Earth. [...] the model systems of science often work similarly to these familiar fictions. (2006, p. 735)

On this view, then, models are akin to the places and characters in literary fiction. When modelling the solar system as consisting of ten perfectly spherical spinning tops physicists describe (and *take themselves* to describe) an imaginary physical system. Likewise, when considering an ecosystem with only one species and when investigating an economy without money and transaction costs, scientists describe imaginary objects and scenarios that are like the characters in works of fiction.

In the last section we said that the notion of fiction that is relevant for the fiction view of models is the second notion of fiction, fiction as imagination. Model-descriptions, then, are prescriptions to imagine something. We also said that, per se, this does not imply that model-descriptions are false because one can be mandated to imagine false and true things alike.

A critic might try to push back and complain that, our assurances notwithstanding, Godfrey-Smith's examples show that what is at work here is the first notion of fiction, fiction as infidelity, because Sherlock Holmes' London and Middle Earth don't exist. This is a misapprehension. First, Sherlock Holmes' London is like real Victorian London, just with an imaginary detective inserted into it, and what Conan Doyle's novels say about London is mostly true. Middle Earth is inspired to a large extent by medieval Europe and much of what Tolkien's novels say is true of medieval European places. So it's simply not the case that the texts of these novels are wholesale fictions in the sense of being falsities. To be sure, some parts of them are falsities: Sherlock Holmes is not a real person, and there are no hobbits, not even in a medieval context. But we can't infer from the fact that a statement occurs in a work of fiction that the statement is false. Second, and probably more importantly, whether or not the characters described in a work of fiction exist does not matter for a productive engagement with the novel. If it turned out that Conan Doyle based his stories on a real detective and that there actually was a description-fitting person, this would not change the way in which we engage with the novel. And cases of this kind occur. Fan communities are often keen to find out where an author's

---

(2014). For discussions of fictionalism beyond philosophy of science see the contributions to Kalderon's (2005a) for fictionalism in metaphysics, Kalderon's (2005b) for fictionalism in moral philosophy, and Balaguer's (2018), Contessa's (2016), and Leng's (2010) for fictionalism in mathematics.

6.3 Fiction and the Problem of Carriers

inspirations come from, and in doing so find real-world persons on which fictional character are based. When it was discovered that J. K. Rowling's Severus Snape was in fact John Nettleship, the head of science at Rowling's old school, this made headlines across the country. It was reported that Nettleship was initially taken aback by the revelation and that it took him a while to come to terms with his unintentional stardom. All this may be interesting in many ways, not least to Nettleship himself, but it makes no difference to readers of the Harry Potter books, who mostly will not know Nettleship and, indeed, have no knowledge of him.

Models are like fictions in that they canvass an imaginary scenario that the scientists are invited to explore. As discussed previously, whether or not the elements of these scenarios have real-world correspondents, or whether or not the scenarios relate to parts or aspects of the world in an interesting and specifiable way is, at this point, simply bracketed. In fact, leaving open the question how, if at all, the model scenario relates to the real world is the defining feature of the first branch of the fiction view. Positions belonging to this branch insist that all that one gets out of the analogy between models and fiction is an understanding of what models are and insist that an analysis of the model-world relation, and indeed a discussion of other issues pertaining to representation, have to be conducted outside this analogy. This squares with scientific practice, where, when a model is proposed, it is often unknown whether the entities in a model have real world correspondents, and if they do what the nature of the correspondence is. Consider the example of the Higgs boson. Elementary particle models that featured the Higgs boson have been formulated in the 1960s. Scientists spent a great deal of work on studying these models and on understanding what was true and what was false in them. Yet, this implied no commitment to either the existence or nonexistence of the Higgs boson. In fact, modellers remained expressly non-committal about this question and referred it to their experimental colleagues in CERN, where, after much work, a Higgs particle was found in 2012, and it was agreed to be "the" Higgs boson in 2013. Viewing models as fictions does not prejudge the matter either way. The question whether there is a Higgs boson stands outside the model in which it appears, just as the question whether Severus Snape corresponds to a real person stands outside the Harry Potter novels.

One may now grant the point that viewing models as fictions carries no commitment to their falsity, but still be left wondering what interest there is in drawing an analogy between models and fictions. Chairs are like spacecrafts and protons in that they are material things, and number five is like letter "B" and the proposition that 193 countries are members of the United Nations in that they are abstract objects, but little of any interest seems to follow from these observations. Why is the observation that models are on par with places and characters in works of fiction more than a philosophical inert and ultimately inconsequential truism? In particular, why should philosophers of science be interested in it?

The reason they should is because the fiction view has heuristic power. As we have seen in Sect. 1.5, an account of representation has to address the Problem of Carriers, which has two sub-problems: the Problem of Ontology and the Problem of Handling. The latter covers five issues: identity (when are two models identical?), property attribution (what features or properties does a model have?), comparisons (how do scientists compare models to targets and to other models?), truth conditions (what does it mean for a claim to be true in a model?), and the epistemology of models (how do scientists learn about the features of a model?). The heuristic power of the fiction view of models lies in the realisation that all these issues equally arise in connection with literary fiction and that there is a rich literature on fiction that offers a number of solutions. So one might hope to glean an answer to the Problem of Handling from the literature on fiction. Likewise, one might hope to obtain an answer to the Problem of Ontology from that same literature. The ontology of fiction has been discussed extensively and once a connection between models and fiction is established, philosophers of science can tap into the literature on fiction and avail themselves of the solutions that are already there.

The five questions that fall under the Problem of Handling come up in the context of fiction as follows. If confronted with two texts, there is a question whether they describe the same fiction. A good translation is such that it describes the same fictional world as the original text; a bad translation doesn't. On what grounds can it be decided which is the case? This is an instance of the problem of identity conditions. We also ascribe properties to fictional characters in seemingly much the same way in which we ascribe properties to actual objects. We say that Hercule Poirot is Belgian with the same ease with which we say that Gerhard Richter is German. Likewise, we engage in comparisons, for instance when we say that Ken McCallum (the current director general of MI5) is less eccentric than Sherlock Holmes, and that Poirot's mannerisms are more irritating that Holmes'. We can truly say that Poirot is a detective while it is false the he's a yoga teacher. We can also truly say that Poirot has a heart and a liver, and it is false that his body temperature is 15 degrees Celsius. Only the first of these claims is an explicit part of the content of Agatha Christie's novels, yet there are matters of fact about Poirot's anatomy and about his yoga habits "in the world of the story", even when claims concerning these matters go beyond what is explicitly stated in the text. It is furthermore crucial that these facts are accessible to us. We can find out whether or not certain things are the case through engaging with the novel, even if they are not explicitly stated – readers are able, as it were, to fill the gaps in a story.[12]

With the parallels between models and fiction mapped out, one can now turn to the literature on fiction and look for answers to our problems. This is where the real work begins. As noted, the analogy between models and fiction neither settles the status of models, nor does it solve problems. What it does is make options available, and there is a plethora of positions to choose from.[13]

---

[12] The parallel between models and fictions is further discussed in Friend's (2020), Frigg's (2010b), Godfrey-Smith's (2006, 2009), Levy's (2012, 2015), Thomson-Jones' (2010), and Toon's (2010).

[13] For a surveys and discussions of the options available see Crittenden's (1991), Friend's (2007), Kroon and Voltolini's (2018), and Salis' (2013). Woodward (2011) provides survey of theories of truth in fiction. For a recent discussion of scientific fictions in the framework of possible worlds see Gallais' (2019).

6.3 Fiction and the Problem of Carriers

As we have seen at the beginning of this chapter, scientists (often) engage in the face value practice, the practice of talking and thinking about models as if they were real. A seemingly natural way to make sense of the face value practice is to take it at face value. On such a reading, scientists not only talk about models as if they were real; in fact, they are real. This is the position of *fictional realism*. According to fictional realism, first appearances notwithstanding, there *are* fictional entities. So when scientists talk about the ideal pendulum or an isolated population, they talk about real entities and when they investigate a model they discover genuine features of these entities. Rather than being "descriptions of missing systems", there is something that model-descriptions correctly describe. The "systems" of these descriptions are not missing, and we were misled into believing that they are missing by the fact that the objects that fit these descriptions typically don't exist as ordinary physical objects. But there are more objects in the world than just ordinary physical objects, and the objects that match the descriptions are abstract objects. *Fictional antirealism* denies this and submits that there are no such things as fictional entities and the "as if" in the face value practice ought to be taken seriously. Antirealists therefore insist that talk of fictional entities ought not to be taken literally and that reference to fictional entities is only apparent and can be paraphrased away.

The pioneer of fictional realism was Alexis Meinong (1904), who thought that in addition to ordinary physical objects that exist, "*there are* – in an ontologically committed sense – things that do not exist" (Friend 2007, p. 141). This position has become known as *Meinongianism*, and "neo-Meinongean" positions have more recently been formulated by Parsons (1980) and Zalta (1983), among others. *Possibilism* takes an altogether different route and regards fictional objects as ordinary objects that just happen to exist not in the actual world but in another possible world (Lewis 1978). To characterise these fictional entities along one of these lines and to understand their nature is a formidable philosophical task. There have been a few suggestions along those lines, and we will briefly discuss these in Sect. 6.6. However, by and large this is unexplored territory.

There is a question, however, whether an expedition into this territory is worth the effort, at least if the objective is to develop a theory of scientific models. The worry is that the realist may well have overshoot the aim. One begins by lamenting the absence of a description-fitting object, but then one ends up having not one, but an infinity of such objects. Model-descriptions, like the texts of novels, only contain a finite number of specifications and leave all other things open. The Sherlock Holmes stories don't specify whether Holmes could read Arabic transliterations of Tamasheq and whether he appreciated Caravaggio's *Amor Vincit Omnia*, and so there are description-fitting objects that do read Arabic transliterations of Tamasheq and appreciate Caravaggio's painting, and ones that don't.[14] The same goes for any of the (presumably infinitely many) properties that aren't explicitly specified in the

---

[14] In the philosophy of literature this is known as the problem of incompleteness. For a discussion see Howell's (1979).

text, and so there is an infinity of description-fitting objects. The question then is: which of these is the "true" Holmes, and how can we find out? This question is important because we need to know whether claims like "Holmes can read Arabic transliterations of Tamasheq" are true or false in the Holmes stories.

As Friend (2020, Sect. 5) notes, postulating fictional entities does not help us answer these questions. The only thing we can do to discover the properties of Holmes, and to find out which of the infinitely many description-fitting objects is Conan Doyle's Holmes, is to engage with the text of the novel and draw inferences from what is explicitly stated. We may, for instance, go through the stories with a fine-toothed comb, gather everything that Conan Doyle says about Holmes' artistic tastes, and based on this information come to a view about Holmes' appreciation of Caravaggio's painting. The same goes for modelling. Postulating description-fitting systems for, say, the model-description of Newton's model leaves us with an infinity of different model systems, and to find out which one is the "true" system one has to draw inferences from the given description. The question is how this is done. But this question is not answered by taking the plunge and postulating fictional objects, but rather by formulating an account of how readers engage with a fictional text, or, indeed, scientists with a model-description.

According to Friend (ibid., pp. 103–04), there are two general lessons for the philosophy of science in this. First, the parallel between models and fiction does not settle the question of the ontology of models, indeed it does not even narrow down the available options. Second, if our concern is to understand modelling and scientific representation, the focus on the ontological status of model-systems is misguided. Students of models can deal with questions about the use of models in scientific practice and their representational function without settling the ontological status of models, which they can simply bypass and leave to metaphysicians. Instead, philosophers of science should focus on developing an account of how scientist reason with models and on how they draw inferences from model-descriptions. In our terminology this amounts to saying that we should only deal with half of the Problem of Carriers: address the Problem of Handling and set aside the Problem of Ontology.

Friend is right in pointing out that much of what matters about models from a philosophy of science perspective does not depend on the ontology of carriers and that those interested in scientific representation would be ill-advised to get bogged down in the metaphysics of fictional entities. However, as we will see in Sect. 6.6, the Problem of Ontology is not completely irrelevant for the issue of representation and those concerned with representation need at least make some commitments concerning ontology. With this in mind, we now turn to the discussion of the handling of carriers. Recent discussions of the fiction view of models have predominantly focussed on a framework that is known as pretence theory, which gained prominence through Walton's (1990). Friend commends this focus and sees this as the right approach to address the questions that arise in the practice of model-based science. For these reasons we briefly review the main tenets of pretence theory in the next section and then discuss how this theory can be used to address the Problem of Handling.

6.4 A Primer on Pretence Theory

Pretence theory takes as its point of departure the capacity of humans to imagine things.[15] Sometimes we imagine something for no particular reason. But there are cases in which our imagining something is prompted by the presence of a particular object, in which case this object is referred to as a *prop*. An object becomes a prop due to the imposition of a *principle of generation*, prescribing what is to be imagined as a function of the presence of the object. If someone imagines something because they are prompted to do so by the presence of a prop, they are engaged in a game of make-believe. Someone who is involved in a game of make-believe is *pretending*. To describe someone as pretending is a shorthand for saying that they are participating in a game of make-believe and has, in this context, nothing to do with deception. Pretence theory is the systematic study of games of make-believe. The simplest examples of games of make-believe are children's games. In one such game stumps may be regarded as bears and a rope put around the stump may mean that the bear has been caught. In the context of this game, little Rosa is prescribed to imagine a bear when she sees a stump and to imagine that she caught the bear when she puts a rope around the stump.

Pretence theory considers a vast variety of different props ranging from stumps to novels, and from paintings to theatre plays. For our current purposes we only consider the case of literature. Works of literary fiction are, from the current point of view, props because their purpose is to prompt the reader to imagine certain things. By doing so, fictions generate their own games of make-believe. Such a game can be played by a single player when reading the work, or by a group when someone tells the story to the others.

Some principles of generation are ad hoc. Regarding stumps as bears, for instance, is a principle that a group of children can impose spontaneously, and the principle can change quickly. Other principles are publicly agreed upon and relatively stable. Games based on public principles are *authorised*; games involving ad hoc principles are *unauthorized*.[16]

Props generate *fictional truths* in virtue of their own features and the principles of generation. Consider a work w, for instance Conan Doyle's Holmes stories. Let the w-game of make-believe be the game of make-believe based on w, and similarly for "w-prop" and "w-principles of generation". A proposition p is then fictionally

---

[15] We discuss pretence theory as it is developed in Walton's (1990). Currie (1990) and Evans (1982, Ch. 10) develop different versions. Our summary of pretence theory in this section is based on Frigg's (2010a).

[16] According to Walton, a prop is a *representation* iff it is a prop in an authorised game. On this view, then, stumps are not representations of bears because the rule to regard stumps as bears is an ad hoc rule, while Sherlock Holmes is a representation because everybody who understands English is invited to imagine the content of the novels. It is important to note that this use of "representation" is specific to pretence theory and does not accord with the way the term is used in theories of scientific representation, where it concerns the relation of a carrier to something beyond itself. We will not use Walton's notion of representation in what follows.

true in w iff the w-prop together with the w-principles of generation prescribes p to be imagined in the w-game of make-believe.

In common parlance "imagination" is often associated with an unbridled flow of free thoughts. Not so in pretence theory. Imagination is regimented and regulated, and the notion of fictional truth is both normative and objective because a certain proposition is fictionally true only if the prop together with the principles of generation prescribe it to be imagined. Nothing prevents a reader from imagining Sherlock Holmes as boarding a spacecraft and traveling to another galaxy, but the Conan Doyle stories do not prescribe these imaginings and the proposition that "Sherlock Holmes is boarding a spacecraft and traveling to another galaxy" is not fictionally true in the Conan-Doyle-Holmes game of make-believe.

Being imagined by someone is not a necessary condition for a proposition to be a fictional truth. Fictional propositions are ones for which there is a prescription to the effect that they have to be imagined, and whether a proposition has to be imagined is determined by the prop and the principles of generation. Hence, props, via principles of generation, make propositions fictionally true independently of people's actual imaginings. For this reason there can be fictional truths that no one knows of, and one can discover fictional truths that were hitherto unknown. The "world of a fiction w" is the set of all propositions that are fictionally true in w.

Fictional truths can be generated directly or indirectly. Directly generated truths are *primary* and indirectly generated truths are *implied*. Derivatively, one can call the principles of generation responsible for the generation of primary truths *principles of direct generation* and those responsible for implied truths *principles of indirect generation*. The leading idea is that primary truths follow immediately from the prop, while implied ones result from the application of some rules of inference. The reader of David Lodge's *Changing Places* reads that Zapp "embarked [...] on an ambitious critical project: a series of commentaries on Jane Austen which would work through the whole canon, one novel at a time, saying absolutely everything that could possibly be said about them" and is thereby invited to imagine the direct truth that Morris Zapp is working on such a project. The reader is also invited to imagine that Zapp is overconfident, arrogant in an amusing way, and pursues a project that is impossible to complete. None of this is explicitly stated in the novel. These are inferred truths that the reader deduces from common knowledge about academic projects and the psyche of people pursuing them.[17] What principles can legitimately be used to reach conclusions of this sort is a difficult issue fraught with controversy to which we will return briefly below.

Whether the imagination is true or false is of no concern to the game of make-believe. There could be a real bear just behind the stump, and, as it turns out, Severus Snape has a real-world counterpart. That this is so may be interesting in many ways, but it has no bearing on the fiction as such. According to pretence theory, the

---

[17] The distinction between primary and inferred truths is not always straightforward, in particular when dealing with complex literary fiction. Walton also guards against simply associating primary truths with what is explicitly stated in the text and inferred ones with what follows from them. For our current purposes we can set these complexities aside.

essential difference between a fictional and non-fictional text does not lie in its truth but in what we are supposed to do with the text: a text of fiction invites us to imagine certain things while a report of fact leads us to believe what it says. We can imagine both what is and what is not the case and hence fictional truth is compatible, and may in fact coincide, with actual truth.

What view about the ontology of fictional entities does pretence theory imply? The answer is "none". Walton favoured antirealism and denied that there are fictional entities. Pretence theory is certainly compatible with an antirealist account of fiction because no ontological commitments are incurred when playing a game of make-believe. However, pretence theory neither entails nor presupposes antirealism. As we will see in Sect. 6.6, realists about fictional entities can also appeal to pretence to account for a number of features in our practice of dealing with fiction. This is the reason why Friend, as noted at the end of Sect. 6.3, recommends (in our terminology) that philosophers of science bypass Problem of Ontology and focus on the Problem of Handling, to which pretence theory offers a promising response.

6.5 Models and Make-Believe

Models are usually presented to us by way of descriptions.[18] The core idea of an approach that aims to use pretence theory to respond to the Problem of Handling is to understand these descriptions as props in a game of make-believe in much the same way in which the text of novel is a prop in a game of make-believe.[19] This squares with the practice of modelling where model-descriptions often begin with "consider" or "assume", which make explicit that the description to follow is not a direct description of a target but prescription to imagine a particular situation.[20] Although it is often understood that the situation is such that it does not occur anywhere in reality, this is not a prerequisite. Models, like literary fictions, are not *defined* in contrast to truth (recall our discussion in Sect. 6.2). In elementary particle physics, for instance, a scenario is often proposed simply as a suggestion worth considering and only later, when all the details are worked out, the question is asked whether this scenario bears an interesting relation to what happens in nature, and if so what the relation is.[21]

---

[18] We here focus on verbal model-description for ease of presentation. There is no assumption that model-descriptions must be verbal and in principle models can be specified through any means we like, for instance through drawings.

[19] If the models are presented through other means, for instance through an image or a graph, these other means can be taken to be the prop.

[20] This approach to the Problem of Handling is developed in Frigg's (2010a, b). A related approach is introduced in Barberousse and Ludwig's (2009).

[21] For an accessible account of particle physics that makes this aspect explicit see Smolin's (2007, Chap. 5 in particular).

The "working out" of the details of a model consists in deriving conclusions from the basic assumptions of the model and some general principles or laws that are taken to be in operation in the context in which the model is used. For instance, we derive that the planets move in elliptical orbits from the basic assumptions of the Newtonian model and the laws of classical mechanics. This is explained naturally in terms of pretence theory. What is explicitly stated in a model-description are the primary truths of the model, and what follows from them via laws or general principles are the implied truths. The laws and principles that are used in these derivations play the role of principles of generation. So, in the current framework, general laws like Newton's equation of motion play the role of principles of generation that allow those who engage in the game of make-believe to derive secondary truths from primary truths.

This simple idea provides answers to the five questions that fall under the Problem of Handling (Sect. 1.5). Let us begin with *truth in a model*. As we have seen in the previous section, a proposition p is fictionally true in work w iff the w-prop together with the w-principles of generation prescribes p to be imagined in the w-game of make-believe. In the context of modelling, the w-prop is the model-description and the w-principles of generation are the laws of nature (or other general scientific principles) that are taken to be in action in the model. The w-game of make-believe is to imagine what the prop together with the laws of nature taken to be in operation in the context of the model mandate scientists to imagine. Truth in a model is fictional truth in the sense of pretence theory: a claim p is true in the game of make-believe of the model iff the model-description together with the laws of nature assumed to be at work in the model mandate a scientist to imagine p. It is true in the Newtonian model that planets move in elliptical orbits because the model-description and the Newtonian law of motion (which is the principle of generation in this context) mandate scientists to imagine that they do; it is false in the Newtonian model that planets move in square orbits because the model-description and the Newtonian law of motion do not mandate scientists to imagine that they do. Truth in the model is independent of truth simpliciter. It is true in Thomson's model of the atom that electrons are surrounded by a medium of positive charge, like plums embedded in a pudding. We now know that this claim is false, but it is nevertheless true in the model.

As we have seen in the previous section, that someone actually imagines p is not a necessary condition for a proposition to be a fictional truth. Props, together with principles of generation, make propositions fictionally true independently of people's actual imaginings. So there is a robust sense in which scientists can discover a truth in the model that was hitherto unknown.

This take on truth in a model also provides an *epistemology of modelling*: we investigate a model by finding out what follows from the primary truths together with the principles of generation. This seems to be both plausible and in line with scientific practice because a good deal of the work that scientists do with models can be accurately described as studying the consequences of the basic assumptions of the model.

The world of a fiction w is the set of all propositions that are fictionally true in w. Likewise, the world of a model is the set of all propositions that are true in the model. This gives us a straightforward response to the issue of *identity conditions*: two models are identical iff the worlds of the two models are identical. This condition does *not* say that models are identical if the model-descriptions have the same content. In fact two models with the same model-descriptions (the same prop) can be different because different principles of generation are assumed to be in operation. This is the case, for instance, when one treats what might look like "the same model" first classically, and then quantum mechanically. A common model-description of a model of a hydrogen atom is to say that the model consists of an electron and proton that attract each other with a Coulomb force. If we assume that the laws of classical mechanics serve as the principles of generation in the model, we get the Bohr model and it is true in the model that electrons move in precisely defined trajectories. If we assume that the laws of quantum mechanics serve as the principles of generation in the model, we get the Schrödinger model of the atom and it is false in the model that electrons move in precisely defined trajectories. Regarding these as different models despite being based on the same model-description is the right verdict.

Property attribution is analysed as it being true in the game of make-believe that the model system possesses a certain property. To say that the model-population is isolated from its environment is just like saying that Zapp drives a convertible. Both claims follow from a prop together with principles of generation. So there is nothing mysterious about ascribing concrete properties to nonexistent entities, nor is it a category mistake to do so.

The analysis of *comparative statements* is more involved. The problem is that comparing a model with either another model or a real-world object involves things that are not part of the game of make-believe, and hence are not covered by the framework. How to best overcome this problem is a matter of controversy, and different suggestions have been made. Walton's own view is that we devise an unauthorised game of make-believe to make such comparisons, one that contains the constituents of both models, or the model and the real object, and then carry out comparisons within that extended game of make-believe.[22]

6.6 Make-Believe, Ontology and Directionality

After having introduced model props, model principles of generation, and model games of make-believe one may wonder: but where is the model? In Frigg's original version of the fiction view the model was associated with the imaginings of a scientist in the game of make-believe (2010a, p. 266): model-descriptions are props that prescribe scientists to imagine certain things, and the model is what they imagine.

---

[22] See Walton's (1990, pp. 405–16). Lamarque and Olsen (1994, Chap. 4) solve the problem by introducing fictional characters. See Frigg's (2010a), Godfrey-Smith's (2009), Salis' (2016, 2019) and Thomasson's (2020) for discussion of this problem in the context of models.

This implies a firmly antirealist answer to the Problem of Ontology: there are no carriers – they are a figment of the imagination. This answer is in line with Walton's own account of fictional objects, which is also antirealist.

This view faces a difficulty, however, when paired with an account of how models represent. Many accounts of representation involve the posit that the carrier denote the target. As we have seen in Chaps. 2 and 3, denotation is a dyadic relation that obtains between certain symbols and certain objects. But, as Toon notes (2012a, p. 58), if carriers don't exist they can't denote because only something that exists can enter into a denotation relation with something else.

This leaves the fiction view with two options. The first is to renounce the commitment to "real" denotation and submit that scientific representation only requires *pretend denotation*. When using a carrier to seemingly denote a target, scientists only claim *in pretence* that the model system denotes the target. Statements like "the Newtonian two-body system denotes the Sun-Earth system" therefore must be understood as being implicitly prefixed with a fictional operator, which makes the full claim something like "in the game of make-believe for Newton's model, a scientist uses the two-particle model system to denote the sun-earth system". This statement can be assessed for genuine truth if there is a rule according to which, in the game of make-believe for Newton's model, a scientist can use the two-particle model system to denote the Sun-Earth system. But even if true, models have no real denotation.

Is pretend denotation too weak a notion to figure in a tenable response to the Epistemic Representation Problem? Salis thinks so and presents what she calls *the new fiction view of models*, which reworks the fiction view so that models come out as objects that exist (and that can therefore enter into denotation relations). According to this view, a model should be regarded as a "complex object constituted by a model-description D and content C, so that $M = [D, C]$" (Salis 2019, p. 11). This view retains the original analogy between models and literary fiction, but sees the analogy in a slightly different place: "[f]rom an ontological point of view, the model M is analogous to a literary work of fiction; the model-description D is analogous to the text of a fictional story […] and the model-content C is analogous to the content of a fictional story" (ibid.). The view is designed to strike the right balance between the demands of the substantive theory of representation, which, Salis submits, needs denotation, and the antirealist desire to avoid commitment to fictional entities. The model exists because it consists of a set of sentences and their content, which exist. Yet no commitment to the existence of fictional entities is incurred. In fact, the "new view" is not only compatible with antirealism about fictional entities; as Salis points out it actually implies antirealism about them (ibid.). The crucial difference with the previous approach is that it doesn't regard models as fictional entities; rather it regards models as descriptions and their content, and hence antirealism about fictional entities is compatible with realism about models.[23]

---

[23] For a further discussion of this view see Salis et al. (2020).

An alternative way of reframing carriers so that they exist and can enter into denotation relations is to opt for fictional realism (see Sect. 6.3). Contessa (2010, 223–28) submits that models are abstract objects that stand for a set of possible concrete systems, but without further elucidating the nature of abstract objects. Thomasson (1999) developed a version of fictional realism according to which fictional entities are neither Meinogian objects nor possiblibila, but what she calls *abstract artefacts*. The label reflects the fact that on her account fictional characters, even though they are abstract objects, are brought into existence by a creative act of a writer just like a vase is brought into existence through the creative act of the potter. The view agrees with the pretence account that discourse about the content of a story occurs within pretence, and that pretence is such that there may be no description-fitting objects in the physical world. At the same time, and this is where abstract artefactualism diverges from the Waltonian account of pretence, Thomasson insists that when writing a story and introducing a game of make-believe, an author creates an abstract cultural artefact. Such artefacts can be talked about from an "external" perspective like physical objects. We say that Sherlock Holmes was created by Conan Doyle in the same way in which we say that *The Last Supper* was painted by Leonardo da Vinci, and both claims are true *simpliciter*. Artefactualism thus preserves the advantages of pretence theory in accounting for discourse that is internal to fiction, while at the same time regarding fictional characters as things that exist. Thomasson (2020) and Thomson-Jones (2020) both note that while this approach has gained traction in the literature on the metaphysis of fiction, it has been overlooked in the literature on scientific models, and they argue that carriers are best analysed as abstract artefacts. The suggestion offers the same solution to the Problem of Handling as the pretence account of models, while, at the same time, meeting the requirement that carriers have to be the sort of objects that can enter into denotation relations.[24]

Both Salis' new fiction view and Thomasson's and Thomson-Jones' artefactual view provide workable suggestions of the ontology of carriers. In their current formulation they are, however, not integrated into an account of representation and, going forward, the main challenge will be to combine them with such an account in a way that answers the full set of problems concerning scientific, or epistemic, representation.[25]

---

[24] Knuuttila (2017) also draws on the idea that models are artefacts, but she emphasises the primacy of the model-descriptions themselves, and indeed the particular ways in which they are presented – ink on paper, markers on whiteboards and so on – as part of the model. So Knuuttila's artefactualism is prima facie different from Thomasson's and Thomson-Jones'. For a brief discussion of the relationship see Knuuttila's (2017, Sect. 5.1).

[25] At the time of writing Salis' account was too recent to have elicited responses. For a first critical discussions of the artefactualist approach see Friend's (2020) and Godfrey-Smith's (2020).

6.7 Direct Representation

At the beginning of this chapter we divided fiction-based accounts of models into two branches, and we have so far only dealt with the first branch. The time has come to turn to the second branch. Positions belonging to that branch use the parallel between models and fiction to provide a full-fledged account of representation. This widens the scope of the discussion to include the full set of problems formulated in Chap. 1. As noted previously, authors in the second branch also make use of pretence theory, and so the machinery we have put into place in Sect. 6.4 will now be used again, albeit in a rather different way.

The theories of representation we have encountered so far posit that there are model objects of some sort and construe epistemic representation as a relation between two entities, the carrier and the target system. As we have seen in Sect. 3.1, Weisberg calls this indirect representation and sees it as the defining feature of modelling. Levy, speaking specifically about the fiction view of models, refers to it as the *whole-cloth fiction view* (2012, p. 741). This view contrasts with what Toon (2012a, p. 43) and Levy (2015, p. 790) call a *direct view* of representation.[26] This view does not recognise carriers and instead aims to explain epistemic representation as a form of direct description. Model-descriptions like the description of an ideal pendulum provide an "imaginative description of real things" (Levy 2012, p. 741) such as actual pendula. They are not, as Thomson-Jones maintains, descriptions of a missing system: they describe real targets. Scientific representation therefore does not involve the introduction of model systems of which the pendulum description is literally true (Toon 2010, p. 87; 2012a, pp. 43–44). This approach is now known as *direct representation*. This amounts to a rejection of the main taxonomic distinction that has been made in both the similarity view and the structuralist view. Levy and Toon don't comment on the issue of taxonomy beyond the rejection of indirect representation, and, following Walton, consider a wide array or props ranging from literary texts to statues and geographical maps. This would suggest that they demarcate neither taxonomically nor scientifically, which means that when formulating an account of representation they address the ER-Problem.

Both Levy and Toon develop their account of direct representation using Walton's account that we introduced in Sect. 6.4. In the current context two kinds of props are particularly important. The first are objects like statues. Consider a statue showing Napoleon on horseback (Toon 2012a, p. 37). The statue is the prop, and the game of make-believe for it is governed by certain principles of generation that apply to statues of this kind. So when seeing the statue we are mandated to imagine, for instance, that Napoleon has a certain physiognomy and certain facial expressions. We are not mandated to imagine that Napoleon is, or was, made of bronze, or that he hasn't moved for more than 100 years. The second relevant kind of prop is works of literary fiction. In this case the text is the prop, which, together with principles of generation appropriate for literary fictions of a certain kind, generates fictional

---

[26] Levy (2012, p. 741) earlier also referred to it as the *worldly fiction view*.

6.7 Direct Representation

truths by prescribing readers to imagine certain things. For instance, when reading *The War of the Worlds* (ibid., p. 39) we are prescribed to imagine that the dome of St Paul's Cathedral has been attacked by aliens and now has a gaping hole on its western side.

The crucial move now is to say that models function in the same way. Material models such as an architectural model of the Shard in London are like the statue of Napoleon (ibid., p. 37): the model is the prop and the building is the object of the representation. Non-material models are like the text of *The War of the Worlds*: they are descriptions that mandate the reader to imagine certain things about the system (ibid., pp. 39–40). A model of the ideal pendulum, for instance, is a model-description that prescribes us to imagine that the target, the real ball and spring system we have in front of us, is exactly as the text presents it: we have to imagine the spring as perfectly elastic and the bob as a point mass. Hence, model-descriptions represent actual material objects: the Shard and the bob on a spring. They are not descriptions of missing systems and there is no "intermediary entity" of which model-descriptions are literally true and which, in turn, represents the target. Model-descriptions prescribe imaginings that are directly about a real world target:

Direct Representation: a carrier X is an epistemic representation of a target T iff X functions as a prop in a game of make-believe prescribing imaginings about T.

Direct Representation offers a straightforward answer to the Problem of Carriers. Ontologically speaking, for material models, carriers are physical objects, and for non-material models, they are descriptions, and the Problem of Handing is answered by appeal to pretence theory. The view also offers an explanation of the asymmetry of epistemic representation because the imaginative processes in a game of make-believe are clearly directed towards the target. An appeal to imagination also solves the Problem of Misrepresentation because there is no expectation that imaginations are correct when interpreted as statements about the target. Neither Toon nor Levy address the Problem of Style and remain silent about the applicability of mathematics. These are, however, not problematic omissions and considerations concerning style and applicability could be added to the account along similar lines as in other accounts.

What weighs heavier is that it is unclear how Direct Representation meets the Surrogative Reasoning Condition. Imagining that the target has a certain feature does not tell us how the imagined feature relates to the properties the target actually has (or is at least hypothesised to have), and there is no mechanism to transfer model results to the target. Imagining the pendulum bob to be a point mass tells us nothing about which, if any, claims about point masses are also true of the real bob. One can imagine almost anything about almost any object, but unless there is criterion telling us which of these imaginings should also be regarded as true of the target, these imaginings don't licence surrogative reasoning. Toon touches on this problem and responds:

> We make an inference from the model to the system when we take what is fictional in the model to be true of the system (or, perhaps, approximately true). For example, after some calculations we discovered that it is fictional in our model that the bob's period of oscillation

is $T = 2\pi\sqrt{m/k}$. If we think our model is accurate in this respect, then we will infer that this is not only fictional, but also true of the spring itself. (*ibid*., p. 67, cf. pp. 68–69)

So the mechanism for surrogative reasoning is that the target has the same (or perhaps approximately the same) properties as the model. This also provides us with an account of accuracy: "put simply, a model is accurate in a certain respect if and only if what it prescribes us to imagine in that respect is true of the object it represents" (ibid.).

Poznic (2016a, p. 212) notes that this account is incomplete because it offers no mechanism to distinguish between model claims that are true and ones that are false. And not all false claims should be discarded as irrelevant for our understanding of the model–world relation. As noted previously, models rarely, if ever, represent targets as they are, not even approximately. Many model-properties are imputed to targets only after having undergone transformations. Toon acknowledges this *en passant* when he notices that there are cases in which "similarity seems to play no role at all" (ibid., p. 69). But Direct Representation gives no clue as to what transfer mechanisms that are not based on identity, or similarity, would be (Frigg and Nguyen 2016b, pp. 232–35).

Levy takes a different route. In his (2012, p. 744) he proposed that the Surrogative Reasoning Condition can be met by thinking about scientific representation in analogy with metaphors, but immediately adds that this was only a beginning which requires substantial elaboration. In his (2015, pp. 792–96) he takes a different route and appeals to Yablo's (2014) theory of partial truth.[27] The core idea of this view is that a statement is partially true "if it is true when evaluated only relative to a subset of the circumstances that make up its subject matter – the subset corresponding to the relevant content-part" (Levy 2015, p. 792). The ideal gas model, for instance, prescribes us to imagine all kind of things we know full well to be false (for instance that gas molecules don't collide) and yet the model "is partially true and partially untrue: true with respect to the role of energy distribution, but false with respect to the role of collisions" (ibid., p. 793).

This is a step forward, but it does not take us all the way. Levy himself admits that there are cases that don't fit the mould of partial truth (ibid., p. 794). Such cases often are ones in which the distortive idealisations affect all of the claims the model makes about the target system. For example, a Newtonian model of an object sliding down a frictionless plane is, strictly speaking, entirely false/inaccurate with respect to its subject matter: its velocity as it slides down the plane. In such cases one cannot bracket the distortion – the claim that the plane is frictionless – whilst leaving something that is partially true, since the desired prediction concerning the object's velocity is still false, in virtue of relying on the distortion. Cases of this kind require a different treatment, and at this point it's an open question what this treatment

---

[27] See also Yablo's (2020).

6.7 Direct Representation

would be.[28] These kinds of idealisations are ubiquitous in physics and play an important role in other sciences too, and hence Direct Representation remains incomplete until it has a means to deal with such cases.

The next problem for Direct Representation is that not all models have a target system, which is a serious problem for a view that analyses representation in terms of imagining something *about* a target. Toon addresses this problem by drawing another analogy with fiction. He points out that not all novels are like *The War of the Worlds*, which has an object, namely St Paul's Cathedral. Passages from *Dracula*, for instance, "do not represent any actual, concrete object but are instead about fictional characters" (2012a, p. 54). Models without a target are like passages from *Dracula*. Hence, his solution to the problem is to separate the two cases. When a model has a target then it represents that target by prescribing imaginings about the target; if a model has no target it prescribes imaginings about a fictional character. Toon concedes that models without targets "give rise to all the usual problems with fictional characters" (ibid.). However, he seems to think that this is a problem we can live with because the more important case is the one where models do have a target, and his account offers a neat solution there.

This bifurcation of imaginative activities raises questions. The first is whether the bifurcation squares with the face value practice. Toon's presentation would suggest that the imaginative practices involved in models with targets are very different from the ones involved in models without them, and that these practices require different analyses because imagining something about an existing object is different from imagining something about a fictional character. This, however, does not seem to sit well with scientific practice. In some cases we are mistaken: we think that the target exists but then find out that it doesn't (as in the case of the ether). But does that make a difference to the imaginative engagement with an ether model? Even today we can understand and use such models in much the same way as its protagonists did, and knowing that there is no target seems to make little, if any, difference to our imaginative engagement with the model. Of course the presence or absence of a target matters to many other issues, most notably surrogative reasoning (there is nothing to reason about if there is no target!), but it seems to have little importance for how we imaginatively engage with the scenario presented to us in a model. And such cases are not restricted to science. When visiting the British Museum you can see a bust of Homer. It shows Homer as a blind old man. On Toon's analysis we are supposed to imagine Homer as we see him in the bust, which would be taken to function like the statue of Napoleon. A look at the plaque next to the bust casts doubt on this. There we read that "some question his [Homer's] existence and think that the poems arose from a long tradition of storytelling by many poets". But if it's unclear whether there was a person, Homer, who is the author of the Iliad and the Odyssey, we can't imagine Homer as being in this way or that way. But this does not

---

[28] We discuss this example in some detail in Sect. 9.3. Rice (2018, 2019) also discusses problems with the idea that models can be "decomposed" into the contributions made by their accurate and inaccurate parts.

matter. As in the case of the ether model, the existence or non-existence of target seems to have little, if any, bearing on our imaginative engagement with the bust.

In other cases the opposite happens: a model that was thought not to have a target is found to have one after all. This occurred in the case of Dirac's electron model, which indicated that there were electrons with a "wrong" charge. The existence of "wrongly" charged electrons was initially deemed impossible. The model was deemed flawed and as not having corresponding entities in reality. However, it became clear later that this was a mistake and that there indeed were such entities, namely what we now call positrons. But the existence or non-existence of target entities that did not matter to how scientists imaginatively engaged with the model.

In yet other cases it is simply left open whether there is target when the model is developed. In elementary particle physics, for instance, a scenario is often proposed as a suggestion worth considering and only later, when all the details are worked out, the question is asked whether the constituents of the model correspond to something in the real world, and, if so, what the relation between model entities and their real world correlates is. In fact, much of the work that's done in CERN is concerned with finding out whether the particles that are described in a model exist. So, again, the question of whether there is or isn't a target seems to have little, if any, influence on the imaginative engagement of physicists with scenarios in the research process. In as far as a model is an act of the imagination, nothing in that act changes when targets come and go. Models cross the border from targetless to targeted (and back) unaltered, or they happily stay in the buffer zone between the two.

Levy (2015) offers a different and more radical solution to the problem of models without targets: there aren't any! He first broadens the notion of a target system, allowing for models that are only loosely connected to targets (ibid., pp. 796–97). To this end he appeals to Godfrey-Smith's notion of "hub-and-spoke" cases: families of models where only some have a target (which makes them the hub models) and the others are connected to them via conceptual links (spokes) but don't have a specific target. As an example consider the n-sex population models we introduced in Sect. 1.3. Two-sex population models would be the hub cases, because they do, at least in principle, represent real targets in relevant respects. Four-sex populations would be the spoke cases because they are not models of concrete targets, but they are connected to two-sex models through conceptual links. Levy points out that hub-and-spoke cases should be understood as having a *generalised target*. If something that looks like a model doesn't meet the requirement of having even a generalised target, then it's not a model at all. Levy mentions structures like the game of life and observes that they are "bits of mathematics" rather than models (ibid., p. 797). This is supposed to eliminate the need for fictional characters in the case of targetless models.

This is a heroic act of liberation, but questions about it remain. The core idea of Direct Representation is that a model is an act of imagining something about a concrete object. However, generalised targets such as population growth are not concrete things, and often not even classes of such things. But one cannot reap the ontological benefits of a view that analyses modelling in terms of imaginings about concrete things and at the same time introduce targets that are no longer concrete!

6.8 Against Fiction

Furthermore, the claim that models without targets are "just mathematics" does not come out looking very natural when we look back at the above examples. Neither Maxwell's model of the ether nor elementary particle models that fail to describe existing particles are just mathematics.

6.8 Against Fiction

The criticisms we have encountered above were intrinsic criticisms of particular versions of the fiction view, and as such they presuppose a constructive engagement with the view's point of departure. Some critics think that any such engagement is unnecessary because the view got started on the wrong foot.

A frequent complaint is that regarding models as fictions misconstrues the epistemic standing of models. Portides submits that to label something a fiction is to draw a contrast with something being a truth and therefore "we label X fictional in order to accentuate the fact that the claim made by X is in conflict with what we observe the state of the world to be" (2014, p. 76). For this reason we only classify X is a fiction "if we think that the truth valuation of the claim 'X represents (an aspect of) the world' is false", and classifying a model in this way, Portides argues, "obscures the epistemic role of models" (ibid.). Teller notes that "[t]he idea that science often purveys no more than fictional accounts is very misleading" because even though it has elements that are fictional, the presence of such elements "does not compromise the ways in which science provides broadly veridical accounts of the world" (2009, p. 235). Winsberg says that "[n]ot everything ...] that diverges from reality, or from our best accounts of reality, is a fiction", which is why "we ought to count as nonfictional many representations in science that fail to represent exactly" (2009, pp. 179–80). Among those nonfictional representations we find models like the "frictionless plane, the simple pendulum, and the point particle" (ibid.). The point that Teller and Winsberg make is that even though models often have fictional elements, it would be a mistake to count the whole model as a fiction. In a similar fashion Morrison argues that calling all models fictions is too coarse and that a finer grained distinction should be introduced, namely "one that uses the notion of fictional representation to refer only to the kind of models that fall into the category occupied by, for example, Maxwell's ether models" (2015, p. 90). She characterises the ether model as one "that involves a concrete, physical representation but one that could never be instantiated by any physical system" (ibid., p. 85). Hence "[f]ictional models are deliberately constructed imaginary accounts whose physical similarity to the target system is sometimes rather obscure" (ibid., p. 90).

These criticisms have in common that they characterise fictions as falsities, and hence claim the fiction view of models has to regard models as falsities. This misses the mark. Arguments against the fiction based on a notion of fiction as falsity may be arguments against fictionalism à la Fine, but they have no force against the fiction view of models as developed in this chapter. As we have seen in Sects. 6.2 and 6.3, all versions of the fiction view of models insist that, in the context of the fiction view, fiction must be analysed in terms of imagination and not in terms of falsity. We recognise that the term "fiction" is polysemic, and if something is qualified as a fiction without further comment, then misunderstandings can arise. However, if, after all the qualifications have been provided, someone still insists that the fiction view must be rejected because it brands the products of model-based science as falsities, then this is like attributing to the Standard Model of particle physics, which identifies "quarks" as constituents of matter, the claim that matter consist of a creamy unripened cheese and rejecting it as an absurdity! Proponents of the fiction view are entitled to use "fiction" in the meaning that they find useful, just as particle physicists can define "quark" as they wish.

A related objection is that the fiction view misidentifies the aims of models. Giere deems it "inappropriate" to "regard scientific models as works of fiction" even though they are ontologically on a par (see Sect. 3.4). The reason for this are "their differing functions in practice" (Giere 2009, p. 249).[29] Giere identifies three functional differences (2009, pp. 249–52). First, while fictions are the product of a single author's individual endeavours, scientific models are the result of a public effort because scientists discuss their creations with their colleagues and subject them to public scrutiny. Second, there is a clear distinction between fiction and non-fiction books, and even when a book classified as non-fiction is found to contain false claims, it is not reclassified as fiction. Third, unlike works of fiction, whose prime purpose is to entertain (although some works can also give insight into certain aspects of human life), scientific models are representations of certain aspects of the world. Knuuttila and Magnani make related points. Knuuttila contrast fiction with representation and ask how scientists are supposed to gain knowledge about the world "if scientific models are considered as fictions rather than representations of real-world target systems" (2017, p. 2). Magnani dismisses the fiction view for misconstruing the role of models in the process of scientific discovery. On his account the role of models is to be "weapons" in what he calls "epistemic warfare", a point of view "which sees scientific enterprise as a complicated struggle for rational knowledge in which it is crucial to distinguish epistemic (for example scientific models) from non epistemic (for example fictions, falsities, propaganda, etc.) weapons" (2012, p. 2).

Neither of these objections is on target. As regards Giere's, it's not part of the fiction view to regard "scientific models as works of fiction" in some vague and unqualified sense, much less to claim that literary fictions and scientific models perform the same function. Proponents of the fiction view are careful to specify the respects in which models and fictions are taken to be alike, and none of the aspects Giere mentions are on the proponents' lists. Furthermore, even if they were, they don't drive Giere's point home. First, whether a fiction is the product of an individual or a collective effort has no impact on its status as a fiction; a collectively produced fiction is just a different kind of fiction. Even if *War and Peace* (to take Giere's own example) had been written in a collective effort by all established

---

[29] Similar arguments are made by Liu (2014) and Portides (2014).

6.8 Against Fiction

Russian writers of Tolstoy's time, it would still be a fiction. Vice versa, even if Newton had never discussed his model of the solar system with anybody before publishing it, it would still be science. The history of production is immaterial to the status of a work.

Giere's second point is that there is a clear distinction between fiction and nonfiction books, and he insists that this militates against the fiction view. We find this point somewhat difficult to pin down. Proponents of the fiction view don't say that booksellers should shelve science books containing a model alongside literary fiction. Ultimately Giere's reason for drawing a line between literary fiction and models is that "regarding claims about the fit of scientific models to the world as fictional destroys the well-regarded distinction between science and science fiction" (2009, p. 251). Talk of the "fit" of models to the world suggest that what is stake is models' veridicality. If so, then we're yet again back to the issue of truth and falsity and we keep insisting that the fiction view is not committed to regarding models as falsities.

As regards the third point, proponents of the fiction view agree that it is one of the prime functions of models to represent, and they go to great length to explain how models do this. We have seen in Sect. 6.7 how the direct view analyses representation, and we will see in Chap. 9 how (our version of) the indirect view deals with the issue. Knuuttila's and Magnani's criticisms are also based on an understanding of fiction as falsity, which supposedly implies that fictions can play no epistemic or representational role. We repeat that fiction is not defined through falsity and that models, if understood as fictions in one of the qualified senses discussed in this chapter, *can* play epistemic and representational roles.

Another objection, which is levelled mostly against the indirect view, is that fictions are superfluous and hence should not be regarded as forming part of (let alone *being*) scientific models because we can give a systematic account of how scientific models work without invoking fictions. This point has been made in different ways by Pincock (2012, Chap. 12) and Weisberg (2013, Chap. 4).[30] They argue that scientific models are mathematical objects and that they relate to the world due to the fact that there are mathematical relations between the properties of the model and the properties found in the target system. Weisberg regards fictions as providing a convenient "folk ontology" that may serve as a crutch when thinking about the model, but he takes fictions to be ultimately dispensable when it comes to explaining how models relate to the world.

The view that models are nothing but mathematical structures faces a dilemma. On the one hand, one can take the view at face value. Then one has to accept that models have only mathematical properties, and that representation has to be explained in terms of relations between the mathematical properties of the model and the target. This, at least prima facie, amounts to saying that there is a morphism between the two, which gets us back to the difficulties that we have discussed in Chap. 4. On the other hand, one can try to expand the view and add elements to the

---

[30] Weisberg recognises three kinds of models (2013, Chap. 2): concrete models, mathematical models, and computational models. Of these only the second is relevant for the current discussion.

model that make room for other accounts of representation, but in doing so one invariably ends up with a view that sees models as more than just structures. As we have seen in Sect. 3.4, proponents of the similarity view endeavour to come up with ontologies of models that are rich enough to get the view off the ground, which led to accounts of models in terms of abstract, and indeed fictional, entities. These endeavours involve taking the non-mathematical aspects of models seriously, and so these parts are more than just folk ontology.

Weisberg's preferred solution is to say that a model is a structure plus an interpretation, which he calls a *construal* (2013, p. 39). Construals provide an interpretation of the model structure by assigning variables real world correlates, for instance by saying that variable V is the population density and variable t is time. The first thing to note here is that structures *plus* a construal are more than just structures. Indeed, Weisberg's construals can be seen as Waltonian principles of generation: by assigning variables an interpretation in terms of substantive properties, the construal in effect requires the scientist working with the model to imagine a fictional scenario. The construal of the Lotka-Volterra equation says that the equation is a population-model in which variable V is the prey population density, variable P is predator population density and variable t is time, and to require that these densities are such that the equation so-interpreted correctly describes the time evolution of the variables. But this is tantamount to saying that scientists working with the model are required to imagine a population that has certain properties, which are specified in the construal. So, modulo details that don't matter for the bigger picture, Weisberg's account is in fact equivalent, or at least closely related, to the fiction view as developed in Sect. 6.4.[31]

A different objection to the project of drawing parallels between representation in art and science is that artistic representations have no well-defined target and have little, if any, cognitive value in terms of allowing us to learn about the world. Writing specifically about literary fiction, Currie notes that "[w]e have no more than the vague suggestion that fictions sometimes shed light on aspects of human thought, feeling, decision, and action" (2016, p. 304). So knowledge gained from fiction amounts to little more than truisms. Elgin calls this the argument from banality (2017, p. 245). Since the purpose of scientific models is to give us knowledge, and ideally precise and specific knowledge, they cannot, so the argument goes, be like fictions.

This argument can be resisted in two ways. First, following Elgin (ibid.) one can insist that insights gained from literary fiction are far from truisms, and that fictions can provide extremely fine-grained knowledge. Harper Lee's *To Kill a Mockingbird* provides detailed insights into racial inequality and its dealing with rape, and George Orwell's *Animal Farm* provides a fine-grained account of the phoney pretention of communism. Second, if this was not the case and if one were to agree that the cognitive value of literary fiction is limited, this would not constitute an argument against

---

[31] For a detailed discussion of Weisberg's criticism of fictionalism see Odenbaugh's (2015, pp. 283–87).

6.8 Against Fiction

the fiction view of models. The view is only committed to the abstract mechanism of fiction (as outlined in Sect. 6.4) and it could insist that the implementation of that mechanism in science is relevantly different from its implementation in literary fiction, and that the difference is such that scientific fictions can perform the epistemic functions that literary fictions lack.[32]

Finally, Giere (2009, p. 257) complains that the fiction view plays into the hands of irrationalists. Creationists and other science sceptics will find great comfort, if not powerful rhetorical ammunition, in the fact that philosophers of science say that scientists produce fiction. This, so the argument goes, will be seen as a justification of the view that religious dogma is on par with, or even superior to, scientific knowledge. Hence the fiction view of models undermines the authority of science and fosters the cause of those who wish to replace science with religious or other unscientific worldviews.

Needless to say, we agree that irrationalism must be repudiated. In order not to misidentify the problem it is important to point out that Giere's claim is not that the view itself – or its proponents – support creationism; his worry is that the view can be misused if it falls into the wrong hands. True, but almost anything can. What follows from this is not that the fiction view itself should be abandoned. What follows is that some care is needed when communicating the view to non-specialist audiences. As long as the fiction view is presented carefully and with the necessary qualifications, it is not more dangerous than other ideas, which, when taken out of context, can be put to uses that would (presumably) send shivers down the spines of their progenitors (think, for instance, of the use of Darwinism to justify eugenics).

In sum, if understood in the way intended by its proponents, the fiction view is not open to any of these criticisms.

---

[32] For a discussion of a few of these differences and their effect on cognitive function see Sect. 9.5 and our (2017c, pp. 56–58).

Chapter 7
Representation-As

In this chapter we discuss approaches that depart from Goodman's notion of "representation-as" (Goodman 1976, pp. 27–30). In his account of aesthetic representation the idea is that a work of art both denotes its subject and represents it as being thus or so. A caricature represents a politician as a bulldog; a painting represents its subject as erudite; a statue represents a ballerina as a swan. The general form of representation-as is that a carrier (e.g. a caricature) represents a target or subject (e.g. a politician) as a son-and-so (e.g. a bulldog). The idea that representation-as is not only the modus operandi of many artistic representations, but is also at work in epistemic representations more generally, including scientific models, has been suggested by Hughes (1997; 2010, Chap. 5),[1] and further developed by Elgin (2010; 2017, Chap. 12).[2]

We begin this chapter with a discussion of the demarcation problems in the context of representation-as (Sect. 7.1). We then move on to considering how various authors have used representation-as to answer the ER-problem. Despite talking about representation-as, and acknowledging Goodman's conception of representation-as as a precursor to his views, Hughes' discussion does not take full advantage of the machinery that Goodman and Elgin develop. For this reason, we structure this chapter somewhat anachronistically, starting with Hughes' account (Sect. 7.2), and then outlining Goodman's and Elgin's more developed accounts (Sects. 7.3, 7.4 and

---

[1] Hughes's discussion also influenced both the inferential and interpretational accounts discussed in the Chap. 5. With respect to the former see Suárez's (2004, p. 770), and the similarities and differences are discussed in detail in his (2015). With respect to the latter and see Contessa's (2011, p. 126) for the tentative claim that Hughes' so-called DDI account is a version of the interpretational account. In our (2017a) we discussed Hughes' account in tandem with those accounts. However, as we will see, his reliance on Goodman and Elgin's insights means that the account is more suited to be discussed with theirs.

[2] van Fraassen (1994, 2008) also mentions the notion of representation-as in connection with epistemic representation, although he then proceeds to develop an account more closely in line with those discussed in Chap. 4.

7.5). As we will see, whilst we think that these accounts, Elgin's in particular, are on the right track, they need to be developed in order to provide a full-fledged answer to the ER-problem. Thus, we close by outlining the issues that discussions of representation-as leave unaddressed. This paves the way for our preferred account of epistemic representation in the next chapter (Sect. 7.6).

7.1 Demarcation

As previously, our investigation must start by considering the demarcation problems. Although Goodman's discussion in his *Languages of Art* is primarily concerned with representation in the artistic domain, he considers "symbols" as the subject matter of his discussion, where the class of symbols includes "letters, words, texts, pictures, diagrams, maps, models" (1976, p. xi), and towards the end of the book he explicitly subsumes models under his broader analysis (ibid., pp. 170–73). This suggests that Goodman had little, if any, inclination to demarcate scientifically.

The question whether there is taxonomic demarcation in Goodman would seem to be a matter of interpretation. On the one hand, the project of covering representations as different as texts and pictures under the same analysis might suggest that he would not have demarcated taxonomically. On the other hand, in his discussion of samples and labels he distinguishes between predicates like "red" and samples like red paint-chips (ibid, p. 60), which might suggest that something like the distinction between direct and indirect representation (which we encountered in Sect. 3.1) would also be at work in Goodman's analysis. In this chapter we discuss him as a universalist who doesn't demarcate and hence addresses the ER-Problem. Those who favour the alternative interpretation and attribute to him, and to this followers, a taxonomic distinction between direct and indirect representation can interpret the views we discuss in this chapter as a response to the Indirect Epistemic Representation Problem and simply replace all occurrences of "epistemic representation" (or its cognates) by "indirect epistemic represenation" (or its cognates).

Given the intellectual debt that each of the accounts discussed in this chapter has to Goodman's original discussion of representation-as, it is unsurprising to find that negative responses are given to both demarcation problems by those who invoke representation-as in their discussions of scientific representation. With respect to the Taxonomic Representational Demarcation Problem, Hughes is an adherent of the so-called semantic view of theories, which we discussed in Chap. 4. Therefore, he takes the representational content of a theory to be given by the content of its models: "in any application of the theory it is a local model that represents an individual physical system" (1997, p. 331). This means that there is no demarcation between theories and models. It's also clear he takes his account of representation to apply beyond models: indeed, he introduces it with the example of Galileo's diagrams about the motion of particles (1997, p. 326). This suggests that Hughes doesn't demarcate taxonomically. Elgin, whilst not subscribing to any particular view of scientific theories, treats all scientific representations as being on a par and explicitly includes curve fitting, laws, stylised facts, and idealisations in her analysis (2017, pp. 24–32), as well as models (ibid, pp. 249–72), thought experiments (ibid., pp. 229–35), and laboratory experiments (ibid., pp. 222–28). Furthermore, she discusses these scientific representations alongside visual representations like pictures, diagrams, and maps, which implies that she covers all these representations under the same theoretical umbrella. For these reasons we take neither Hughes nor Elgin to demarcate taxonomically.

With respect to the Scientific Representational Demarcation Problem, we can interpret the fact that both Elgin and Hughes discuss representation-as in scientific models in tandem with representation-as in paintings and other non-scientific representations as implying a negative response to the problem. Hughes writes that:

> Theories [...] are always representations-as, representations of the kind exemplified by Joshua Reynolds' painting of Mrs. Siddons as the Muse of Tragedy. In this painting Reynolds invites us to think of his primary subject, Mrs. Siddons, in terms of another, the Muse of Tragedy, and to allow the connotations of this secondary subject to guide our perception of Mrs. Siddons. Similarly the wave theory of light represents light as a wave motion. It invites us to think of optical phenomena in terms of the propagation of waves, and so to anticipate and explain the behaviour of light. (1997, p. 331)

Moreover, Elgin's discussion in her (1996, Chap. 6; 2010; 2017) makes clear that she sees scientific representation as being analysed in the same way as representation from outside the scientific domain.

For these reasons we will assume that representation-as aims to provide an analysis of epistemic representation in general and therefore has to address the ER-Problem.

7.2 Denotation, Demonstration, and Interpretation

Hughes' account of epistemic representation in terms of representation-as involves three conditions: denotation, demonstration, and interpretation, which is why the account is now commonly known as the DDI account (1997; 2010, Chap. 5). We discuss these conditions in order before turning to how they can be utilised to answer the ER-problem and the remaining problems introduced in Chap. 1.

Quoting directly from Goodman, Hughes proposes that:

> we attend to Nelson Goodman's dictum [reference omitted] that "denotation is the core of representation and is independent of resemblance." Following Goodman, I take a model of a physical system to "be a symbol for it, stand for it, refer to it". (1997, p. 330)

We have already seen the idea that denotation is involved in epistemic representation in Chaps. 2 and 3, but its status is different in the DDI account. While denotation was added to accounts based on stipulation and similarity as a fix for various problems, it is explicitly built into Hughes' account of epistemic representation right from the outset. Denotation is as a two-place relation between a symbol and the object to which it applies. Sometimes denotation is restricted to language: proper names denote their bearers and predicates denote the objects in their extension (sometimes it is restricted even further to only the former). This restriction is neither essential nor helpful. Signs other than words of a certain language can denote. A portrait can denote its subject; a photograph can denote its motif; and scientific model can denote its target system. There is nothing intrinsic in the notion of denotation that would restrict it to language (Elgin 1983, pp. 19–35; 2010, p. 2). This is the first "D" in "DDI". But as discussed in Chap. 2, denotation alone is insufficient for an account of epistemic representation: that X denotes T does not help us understand how X can be used to reason surrogatively about T. Thus, what makes something an epistemic representation, and thereby distinguishes it from a proper name, are the further conditions of demonstration and interpretation.

The demonstration condition, the second "D" in "DDI", requires scientists to establish what is the case in the carrier of a representation. Doing so relies, in Hughes' words, on a carrier being a "secondary subject that has, so to speak, a life of its own. In other words, [a] representation has an internal dynamic whose effects we can examine" (1997, p. 331). The two examples offered by Hughes are both scientific models of what happens when light passes through two nearby slits. One model is mathematical where "[t]he internal dynamic is supplied, at least in part, by the deductive resources of the mathematics they employ" (ibid., p. 332); the other is a physical ripple chamber where they are supplied by "the natural processes involved in the propagation of water waves" (ibid.).

Such demonstrations, either on a mathematical model or on a physical model, or indeed on any carrier, are still about a carrier itself. The final aspect of Hughes' account – the "I" in "DDI" – is the interpretation of what has been demonstrated in the carrier in terms of the target system. This yields the predictions of the model (ibid., p. 333). Unfortunately Hughes has little to say about what it means to interpret a result of a demonstration on a carrier in terms of its target system, and so one has to retreat to an intuitive (and unanalysed) notion of carrying over results from carriers to targets.

With each of the conditions laid out, we can now turn to how they can be put to work in answering the ER-problem. Hughes is explicit that he is not offering denotation, demonstration, and interpretation as individually necessary and jointly sufficient conditions for epistemic representation. He prefers the more "modest suggestion that, if we examine a theoretical model with these three activities in mind, we shall achieve some insight into the kind of representation that it provides" (ibid., p. 339). We are not sure how to interpret Hughes' position in light of this. On one reading, he can be seen as describing how we use epistemic representations in a process of investigation. As such, the conditions function as a *diachronic* account of what a user of an epistemic representation does when utilising a carrier in an attempt to learn about a target system. Thus understood, the view is that the user first establishes that the carrier denotes the target, then proves what we want to know about the carrier's behaviour, and finally "transfers" the results obtained in the carrier back to the target through an interpretation. Details aside, this picture seems correct. The problem is that it does not explain why and how this is possible. Under what conditions is it true that the carrier denotes the target? What kinds of things are carriers that allow for demonstrations? How does interpretation work? That is, how

7.2 Denotation, Demonstration, and Interpretation

can results obtained in the carrier be transferred to the target? These are questions an account of epistemic representation has to address, but which are left unanswered by the DDI account thus interpreted. Accordingly, under the current interpretation, DDI provides an answer to a question distinct from the ER-problem. Although a valuable answer to the question of how models are used diachronically, it does not help us here, since it presupposes the very representational relationship between carriers and their targets that we are interested in.

An alternative reading of Hughes' account emerges when we consider the developments of the structuralist and similarity conceptions discussed in Chaps. 3 and 4, as well as the discussion of deflationism in Chap. 5: perhaps the very act of using a carrier in a way that meets the DDI conditions, with all the user intentions and practices that this brings with it, constitutes epistemic representation itself. And as such, perhaps the DDI conditions could be taken as an answer to the ER-problem:

DDI-ER: a carrier X is an epistemic representation of a target T iff X denotes T and an agent exploits the internal dynamic of X to make demonstrations, which in turn are interpreted by the agent to be about T.

Casting the account in this way throws Hughes' "modest" attitude overboard and regards the three condition as individually necessary and jointly sufficient – this is the price to pay for turning the DDI account into an answer to the ER-Problem.

There is a question here concerning how DDI-ER, as stated, is supposed to capture representation-as, given that the locution doesn't appear in the conditions. Hughes doesn't clarify this, despite his claim that scientific representation is representation-as (Hughes 1997, p. 331). Presumably the idea is that when one uses a carrier X to represent something in a manner meeting the DDI-ER conditions, the user thereby represents T as X. Thus, when one uses a wave model (X) to represent light (T) according to the DDI conditions, one thereby represents light as a wave.

This account comes very close to Interpretation as discussed in the Chap. 5 and as such it serves to answer the questions we set out in Chap. 1 in a similar way. The Surrogative Reasoning Condition is built into the conditions on epistemic representation: by interpreting the result of demonstrating the behaviour of the carrier X in terms of a target T, an agent uses X to reason surrogatively about T. The Problem of Accuracy is answered by the condition that the result of interpreting the behaviour of X in terms of T are true. That the interpretation comes out true is not built into DDI-ER, and thus the account meets the Misrepresentation Condition. Nothing is said about the Problem of Style. However, given the examples that Hughes uses to introduce his account, it's plausible that one could distinguish between different styles in terms of different kinds of carriers allowing for different kinds of demonstrations (mathematical demonstrations of the behaviour of light employ a different style than physical demonstrations in a ripple chamber). Relatedly, since the account is clearly designed to allow that the demonstrations and interpretations can be mathematical, it should meet the Applicability of Mathematics Condition, although, as we discuss below, Hughes' discussion doesn't go into much detail about how this would work.[3] Nothing is said about the Problem of Carriers beyond the fact that they must have an internal dynamic that we can examine. Finally, the denotation condition ensures that the account meets the Directionality Condition.

There are, however, two significant issues with the account as stated. First, since denotation is a two-place relation between a symbol and that which it denotes, and since denotation requires that both the symbol and its denotatum exist, the account cannot in general deal with targetless representations, and thus fails to meet the Targetless Representations Condition. Suárez (2015) argues that the account is substantive rather than deflationary, and this in part motivates why he advocates appealing to "denotative function" rather than denotation (where, presumably, "denotative function" is akin to his representational force; recall the discussion in Sect. 5.2). A carrier can have a "denotative function" if its target is non-actual. This is one way of modifying DDI-ER in such a way that it does allow for targetless models, but fleshing out the account requires explicating what having a denotative function amounts to, and how we should think about these non-actual targets. Another way of thinking about targetless epistemic representations in the context of representation-as is discussed in the following section.

A second issue with DDI-ER is even more significant. The problem is that nothing more is said about what it means to perform a "demonstration" on a carrier or to "interpret" the results of such a demonstration in terms of a target. If the account is to be analysed in the deflationary tradition, as urged by Suárez (2015), then this might not be such a bad thing.[4] However, if so, then the considerations in Sect. 5.3 arise again here. Given that Hughes describes his account as "designedly skeletal [and in need] to be supplemented on a case-by-case basis" (1997, p. 335), one option available is to take the demonstration and interpretation conditions to be abstract and to require that they be filled in each instance, or type of instance, of epistemic representation. As Hughes notes, his examples of the internal dynamics of mathematical and physical models are very different, with the demonstrations of the former utilizing mathematics while the latter rely on physical properties such as the propagation of water waves. Similar remarks apply to the interpretation of these demonstrations (as well as to denotation). But, as with Suárez's account, the definition sheds little light on the problem at hand as long as no concrete realisations of the abstract conditions are discussed. Thus, at least as stated, the account fails to provide enough of an explication of what the demonstration and interpretation conditions amount to.

---

[3] See Bueno and Colyvan's (2011) for further discussions of the account in the context of mathematical scientific modelling.
[4] Suárez (2015) provides further a discussion of the relationship between DDI and Inferentialism.

7.3 Denotation and Z-Representation

Despite Hughes explicitly invoking Goodman's notion of representation-as, he does not take full advantage of the machinery that Goodman, and later Elgin, introduced to explicate it. As we will see, this machinery can illuminate how to account for targetless representations and provide further insight into how we can understand Hughes' demonstration and interpretation conditions. The account of representation-as that we are now going to discuss has been developed by Goodman and Elgin in a string of publications, both joint and single authored. When referring to views shared by both authors, we use the acronym "GE" to refer to them jointly. Thus, analysing GE's discussions of representation-as is our task in this, and the following two sections. In the current section we say more about denotation and introduce the notion of a Z-representation. In the next section we introduce the concept of exemplification. In Sect. 7.5 we see how these can be combined to form the complex referential relation of representation-as.

We have already seen how denotation can be invoked to establish epistemic representation, but GE use it in a more subtle way than any of the previous accounts. For them "denotation is the core of representation" (Goodman 1976, p. 5). And if X denotes T then X is a representation-of T.[5] According to GE, symbols need not just denote single objects; certain symbols may denote multiple objects: "[a] predicate denotes severally the objects in its extension. It does not denote the class that is its extension, but rather each of the members of that class" (Elgin 1983, p. 19; cf. Goodman 1976, p. 19). So the predicate "red" denotes all red things and a model of the hydrogen atom denotes all hydrogen atoms.

Moreover, there can be a number of denotational relationships between a carrier and its target:

> What a picture is said to represent may be denoted by the picture as a whole or by a part of it [...] Consider an ordinary portrait of the Duke and Duchess of Wellington. The picture (as a whole) denotes the couple, and (in part) denotes the Duke. (Goodman 1976, p. 28)

Presumably a part of the picture also denotes the Duchess, another part denotes the Duke's nose, yet another part denotes the Duchess's dress, and so on. In fact, there may, in principle, be an indefinitely large number of denotational relationships that hold between parts of the picture and parts of the situation it denotes. The observation generalises. Whilst a scientific model, as a whole, may denote a target system, parts of the model may also denote parts of the target, and so on for other kinds of epistemic representations. This is not to say that there must be part–part denotational relationships to establish the primary one that holds between the carrier and the situation it denotes. Examples from modern art provide plausible instances where there is only one such relation. We can imagine a uniformly red canvas captioned "Kierkegaard's Mood" which as a whole denotes Kierkegaard's mood (Danto

---

[5] We put systematicity above grammatical correctness when we write "X is a representation-of T".

1981). It's hard to imagine what it would take for a part of the canvas to denote a part of the philosopher's mood. So, whether or not there are such part–part relationships, and how many of them there are, can only be established on a case-by-case basis (and it's worth mentioning here that this aspect of denotation can also be applied to Hughes's DDI account, even though he does not explicitly discuss it).

As discussed in the previous section, any account that takes denotation to be a necessary condition on epistemic representation has difficulty meeting the Targetless Representations Condition. GE handle this by granting that not all epistemic representations represent. Pictures of Pickwick or unicorns do not denote anything simply because Pickwick does not exist and nor do unicorns. Such pictures therefore do not represent anything (Goodman 1976, p. 21). This observation generalises: whenever a carrier portrays something that does not exist then the carrier does not represent anything.

This seems counterintuitive and one is tempted to object: if we recognise a picture as portraying a unicorn, then surely it represents something, namely a unicorn! GE get around this objection by drawing a distinction between "representing" and "being a representation-of". A carrier is an epistemic representation of a target only if the former is a representation-of, that is denotes, the latter. However, a carrier can be a representation without being a representation-of:

> A picture that portrays a griffin, a map that maps the route to Mordor, a chart that records the heights of Hobbits, and a graph that plots the proportion of caloric in different substances are all representations, although they do not represent anything. *To be a representation, a symbol need not itself denote, but it needs to be the sort of symbol that denotes.* Griffin pictures are representations then because they are animal pictures, and some animal pictures denote animals. Middle Earth maps are representations because they are maps and some maps denote real locations. Hobbit height charts are representations because they are charts and some charts denote magnitudes of actual entities. Caloric proportion graphs are representations because they are graphs and some graphs denote relations among real substances. *So whether a symbol is a representation is a question of what kind of symbol it is.* (Elgin 2010, pp. 2–3, emphasis added; cf. Goodman 1976, p. 21)

So whether a carrier is a representation-of depends on whether it denotes something. Whether a carrier is representation depends on whether it belongs to a class of objects that usually denote. In other words, to be a representation something need not denote; but it needs to be an object of the right kind. Thus GE take it to be a mistake to confuse the notions of representation and representation-of:

> What tends to mislead us is that such locutions as "picture of" and "represents" have the appearance of mannerly two-place predicates and can sometimes be so interpreted. But "picture of Pickwick" and "represents a unicorn" are better considered unbreakable one-place predicates, or class terms, like "desk" and "table". [...] *Saying that a picture represents a soandso is thus highly ambiguous between saying that the picture denotes and saying what kind of picture it is.* Some confusion can be avoided if in the latter case we speak rather of a "Pickwick-representing-picture" or a "unicorn-representing-picture" [...] or, for short, of a "Pickwick-picture" or "unicorn-picture" [...] *Obviously a picture cannot, barring equivocation, both represent Pickwick and represent nothing. But a picture maybe of a certain kind – be a Pickwick-picture [...] – without representing anything.* (Goodman 1976, pp. 21–22, emphasis added; cf. Elgin 2010, p. 3)

7.3 Denotation and Z-Representation

This leads to the introduction of the notion of a Z-representation: X is Z-representation if it portrays Z. Phidias's Statue of Zeus at Olympia was a Greek-god-representation without being a representation-of a Greek god; Fra Angelico's *Annunciatory Angel* is an angel-representation without being a representation-of an angel; and (a part of) Paolo Uccello's *Saint George and the Dragon* is a dragon-representation without being a representation-of a dragon; and so on.

The crucial point is that this does not presuppose that X be a representation-of Z; indeed X can be Z-representation without being a representation-of anything. A picture must denote a man to be a representation-of a man. But it need not denote anything to be a man-representation. In fact, the kind of a representation is completely independent of what is denotes: "the denotation of a picture no more determines its kind than the kind of picture determines the denotation. Not every man-picture represents a man, and conversely not every picture that represents a man is a man-picture" (Goodman 1976, p. 26).

If this seems like we're bending words here, the following examples make it plausible that the distinction between a Z-representation and representation-of a Z has manifest analytical value when analysing representations and is not just a bit of academic sophistry to masquerade a problem. To begin with, take your copy of Collin's world atlas off the shelf and open it on the first page. You see a map of the world. The map is a territory-representation because it portrays a territory; it is at the same time also a representation-of a territory, namely the world. Now take a different atlas off the shelf, *The Lands of Ice and Fire* by George R.R. Martin and Jonathan Roberts. It contains maps just like the ones in Collin's world atlas: they show contours of continents, the topography of land, the boundaries of water, they are drawn to a certain scale, and so on. They are territory-representations. But they are not representations-of a territory. They are illustrations of the world according to *Game of Thrones*. But there is no such thing as the world according to *Game of Thrones*, and so these maps do not denote anything, and they are therefore not representations-of anything. Contrast this with the word "world". The word does not portray a territory and is therefore not a territory-representation. It does, however, denote the world, and it is therefore a representation-of the world. Finally consider a caricature by James Gillray from the year 1793.[6] It is a cartoon map showing the British Isles, slight distorted to look like a human torso and topped by the head of King George III. At the bottom of the torso it visibly "farts" battleships down to France. The caricature (with the exception of the inserted head) is still a territory-representation, much like the ones we have just discussed. But it's not a representation-of a territory (at least not primarily); it's a representation of a national attitude. So again, we have an example of something that is a Z-representation without being a representation-of a Z.

Some readers may have been left under the impression that the *sole* purpose of distinguishing between a Z-representation and a representation-of a Z is to cope

---

[6] The caricature can be seen on this website: https://uk.phaidon.com/agenda/design/articles/2015/august/26/propaganda-gets-dirty-in-map/.

with targetless models. This is misapprehension. As the example of the caricature map show, even in cases in which the target exists, the target does not have to be represented by a representation of the same kind. That is, Zs do not always, by some kind of necessity, have to be represented by a Z-representation, even if Zs exist. The fastest dog at the races can be represented by a flash (making it a flash-representation that is a representation-of a dog); in Dutch still life paining a snail-picture denotes humility (making it a snail-representation that is a representation-of humility); and in a Bollywood movie two intertwined roses symbolise that the couple is intimate (making it a two-intertwined-roses-representation that is a representation-of a couple being intimate).

That said, the notion of a Z-representation allows an account of representation to meet the Targetless Representations Condition in an elegant way. When a representation obviously portrays something, namely a Z, but Z does not exist, there is the option to regard it as a Z-representation. To explain how, say, a picture showing a unicorn represents, the account does not have to find something in the world (possibly an abstract object) that the representation represents; it can simply say that it is a Z-representation. Recall Weisberg's n-sex populations, which we encountered in Sect. 1.3. For many of the accounts of representation that we have discussed so far models like these pose a serious problem. Weisberg tries to get around the problem by introducing the hard-to-pin-down notion of a "generalised target" (Weisberg 2013, p. 114), and Levy suggests solving the problem by declaring that there are no targetless models: either a model can be analysed as a "hub and spoke" case, or else it's merely a piece of mathematics (see Sect. 6.7). GE can take a different route and say that n-sex models are n-sex-population-representations which happen not to be representations-of n-sex-populations.

This raises the question of what makes something a Z-representation. How do we classify a picture as a Z-representation? And how does this classification work in cases in which there are no Zs? If there are no griffins, what is the basis for sorting pictures into ones that are griffin-representations and ones that are not? GE analyse Z-representations by introducing the notion of a genre:

> Such an objection supposes that the only basis for classifying representations is by appeal to an antecedent classification of their referents. This is just false. We readily classify pictures as landscapes without any acquaintance with the real estate – if any – that they represent. I suggest that each class of [Z]-representations constitutes a small genre, a genre composed of all and only representations with a common ostensible subject matter [...] And we learn to classify representations as belonging to such genres as we study those representations and the fields of inquiry that devise and deploy them. (Elgin 2010, p. 3)

These genres are habitual ways of classifying and as such they are neither sharp nor historically stable, and they typically resist exact codification (Goodman 1976, p. 23).

Whilst the idea of a genre seems relatively intuitive in the context of pictorial representation, as we will see below in Sect. 7.6, it is difficult to see how this classification works in the context of epistemic representation more generally.[7]

---

[7] Of course, there is more to say how this works in the context of pictorial representation too. See Kulvicki's (2006a, b) for further discussion.

7.4 Exemplification

Putting this issue aside for the moment, we can now turn to the second aspect of GE's account of representation-as: exemplification. An item exemplifies a feature if it at once instantiates the feature and refers to it:[8] "Exemplification is possession plus reference. To have without symbolising is merely to possess, while to symbolise without having is to refer in some other way than by exemplifying" (Goodman 1976, p. 53). An item that exemplifies a feature is an *exemplar* (Elgin 1996, p. 171; Goodman 1976, p. 53). The paradigmatic example of an exemplar is a sample. The swatches of cloth in a tailor's booklet of fabrics (Goodman 1976, p. 53), the chip of paint on a manufacturer's sample card (Elgin 1983, p. 71), and the bottle of shampoo we receive as a promotional gift (ibid.) all refer to relevant features – a pattern, a colour, and a particular hair treatment – and they do so by instantiating them.

Exemplification *requires* instantiation: an item can exemplify a feature only if it instantiates it (Elgin 1996, p. 172). Only something that is red can exemplify redness. But the converse does not hold: not every feature that is instantiated is also exemplified. Exemplification is selective (Elgin 1983, p. 71; 2010, p. 6). An exemplar typically instantiates a host of features but it exemplifies only few of them. Consider the example of a chip of paint on a manufacturer's sample card:

> This particular chip is blue, one-half inch long, one-quarter inch wide, and rectangular in shape. It is the third chip on the left on the top row of a card manufactured in Baltimore on a Tuesday. The chip then instantiates each of these predicates in the previous two sentences, and many others as well. But it clearly isn't a sample of all of them. Under the standard interpretation, it is a sample of "blue", but not of such predicates as "rectangular" and "made in Baltimore". (Elgin 1983, p. 71)[9]

So only selected features are exemplified. But there is nothing in the nature of an object that effects that selection; no features are intrinsically more important than others. Which features are exemplified and which features are merely instantiated is not dictated by the object itself: "nothing in the nature of things makes some features inherently more worthy of selection than others" (Elgin 1996, p. 172). In particular, being conspicuous does not, by itself, turn an instantiated features into an exemplified one. A can of paint spilled on the carpet is a vivid instance of the paint's viscosity and yet it does not exemplify viscosity, or indeed anything else (Elgin 1996, p. 174). Turning an instantiated feature into an exemplified one requires selection. A selection is carried out against certain background assumptions. Someone ignorant of those assumption may be incapable of recognising the selection, and

---

[8] A note on terminology. Following Elgin we use the phrase "exemplified feature" to refer to any property, relation, or pattern that an exemplar instantiates and refers to. There are no restrictions on these features; they can be "static or dynamic, monadic or relational, and may be at any level of generality or abstraction" (Elgin 2017, p. 185). Occasionally GE use the phrase "exemplified property", but it is clear from the context that they are not restricting themselves to one-place properties.

[9] For further examples of selectiveness see Goodman's (1976, pp. 53–54) and Elgin's (1983, pp. 72–73; 2010, p. 5).

"[w]ith a change in background assumptions a symbol can come to exemplify new features" (Elgin 1996, p. 176).

For this reason, converting a "merely instantiated" feature into an exemplified feature through an act of selection is usually done against a relevant background. The same sample card can exemplify rectangularity rather than blueness if used in geometry class; and the paint can exemplify viscosity if poured on a carpet at an industrial fair during a demonstration of the paint's consistency. So exemplification, unlike instantiation, is highly context sensitive.

One is then tempted to ask: what is a context? We doubt that a rigorous definition of a context can be given, but we also doubt that one is needed. For current purposes it is sufficient to think of a context as a certain set of problems and question that are addressed by a group of scientists using certain methodologies while being committed to certain norms (and, possibly, values). These factors determine which of X's epistemically accessible features are representationally relevant.[10]

A crucial feature of exemplars is that they provide epistemic access to the features they exemplify: from an exemplar we can learn about its exemplified features (Elgin 1983, p. 93). This is because they instantiate the features they exemplify in a way that makes them salient. The paint chip makes a particular shade of blue salient and thereby acquaints those using the chip with that shade of blue. An exemplar is therefore not merely an instance of a feature but a *telling instance* (Elgin 1996, p. 17; 2010, p. 5): it presents the exemplified features in a context that is designed to render them salient and make them known to those engaging with the symbol. To be exemplified, a feature not only has to be selected by a context; it also has to be epistemically accessible.[11] We say that a feature that satisfies these criteria is *highlighted*. These considerations can be summarised in the following definition:

Exemplification: carrier X exemplifies feature P in a context C iff X instantiates P and P is highlighted in C, whereby P is *highlighted* in C iff (i) C selects P as a relevant feature, and (ii) P is epistemically accessible in C.

A sample card exemplifies, say, a certain shade of red because the card instantiates it and, in the context of a paint shop, it is selected as relevant and is epistemically accessible (a sample card too small to see with the naked eye does not exemplify red).

Just as parts of a picture can denote parts of its subject whilst the entire picture denotes the subject as a whole, different parts of an exemplar can exemplify different features, all of which may be distinct from those exemplified by the exemplar as a whole. For example, the part of the portrait of the Duke and Duchess of Wellington that denotes the Duke may exemplify ferocity and candour; whilst the part of the

---

[10] Proponents of the similarity view of representation rely on an analogous notion of context to determine which features are relevant for similarity comparisons between models and their targets. See Weisberg's (2013, p. 149) for a discussion.

[11] These aspects of exemplification do not imply each other. A non-epistemically accessible feature may be selected in a context. Elgin (2017, p. 192) memorably cautions against selecting a crocodile's mouth as an exemplar of its colour, and not all epistemically accessible features are selected as relevant.

portrait that denotes the Duchess may exemplify astuteness and wisdom.[12] But, as per our discussion of piecemeal denotation above, whether or not parts of a picture exemplify features in this way depends on the case at hand.

There is a final point to clear up before we turn to using exemplification to define representation-as. So far we have used a realistic idiom to talk about features and their instantiation. This is an expedient that carries no metaphysical commitments. One could provide a nominalist translation for all feature-talk, and the notions of exemplification that drop out can be used in the manner discussed below regardless of the metaphysical position adopted. In fact, GE prefer a nominalist view of properties, and features more generally; see Goodman's (1976, pp. 54–55) for a nominalist formulation of exemplification. For our current purposes nothing hangs on what stance one takes on the question of the metaphysics of features and our discussion remains neutral on the matter.

7.5 Defining Representation-As

We are now in a position to see how the notions of denotation, Z-representation, and exemplification can be combined to deliver an answer to the ER-Problem. Given the locution of representation-as, we are looking for conditions under which X represents T as Z. A first stab would be to say that X represents T as Z if X is a Z-representation and denotes T. This, however, is not yet good enough because it provides no systematic answer to the question of how features of X are transferred to T. A key insight on the way to a definition of representation-as is that Z-representations can, and often do, exemplify features associated with Zs. A caricature that portrays Churchill as a bulldog is a bulldog-picture and it exemplifies bulldog-features like aggressiveness and relentlessness. The crucial step now is to realise that in portraying Churchill in this way, the caricature imputes *these* features to Churchill. Elgin makes this explicit:

> [X] does not merely denote [Y] and happen to be a [Z]-representation. Rather in being a [Z]-representation, [X] exemplifies certain properties and *imputes* those properties or related ones to [Y]. [...] The properties exemplified in the [Z]-representation thus serve as a bridge that connects [X] to [Y]. (2010, p. 10, emphasis added)

This gives a name to the crucial step of "transfer": *imputation*. This step can be analysed in terms of stipulation by a user of a representation. When someone uses X as a representation-as, she has to stipulate that certain features that are exemplified in X be imputed to T.[13]

We emphasise that imputation does not imply truth: T may or may not have the features imputed to it by the representation. So the representation can be seen as

---

[12] The idea that parts of a portrait can exemplify, and therefore instantiate, features like candour or astuteness, is discussed later in the chapter.

[13] Or, in other words, she *ascribes* certain features exemplified in X to Y. In our usage, to impute a feature to the target is the same as to ascribe a feature to the target.

Fig. 7.1 Representation-As

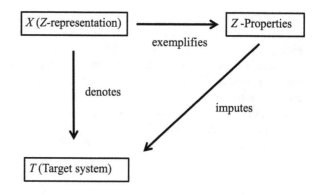

generating a *claim* about T that can be true or false; it should not be understood as producing a truism. Moreover, the claim generated need not be believed by an agent using the representation, thereby allowing for deliberate misrepresentation. Moulding this into an answer to the ER-problem we arrive at the following:

Representation-As: a carrier X is an epistemic representation of T iff X represents T as Z, whereby X represents T as Z iff (i) X denotes T, (ii) X is a Z-representation exemplifying features $Z_1, \ldots, Z_n$, and (iii) at least one of $Z_1, \ldots, Z_n$ is imputed to T.

To make this account more palpable, and to pave the ground for the discussions in Chaps. 8 and 9, Fig. 7.1 offers a graphic portrayal of Representation-As.

To see the account in action, consider the Newtonian model of a planetary orbit that we have seen in the Introduction. The target system is the celestial system, with one of the bodies in the model denoting the sun, and the other denoting the orbiting planet. The model denotes this system. The model is a celestial-orbit-representation that exemplifies the feature of exhibiting gravitational forces between the two bodies and the feature that the object denoting the planet moves in an elliptical orbit around the object denoting the sun. (The model does not, or at least need not, exemplify features like consisting of spheres with homogenous mass distributions.) These exemplified features are then imputed to the actual celestial system. Thus, reasoning about the model, in particular by investigating which features it exemplifies, allows us to generate claims about the system itself. We are now in a position to see how GE's machinery allows us to make more sense of the demonstration and interpretation conditions of DDI-ER. The idea is that the demonstrations discussed by Hughes should be thought of in terms of investigating what features a carrier exemplifies and that the interpretation condition, according to Representation-As, involves imputing those exemplified features to the target.

It's worth briefly clarifying the relationship between Z and the exemplified features stated in (ii). We speak of X representing T as Z, which might suggest that the Z is the crucial ingredient. This is not quite the case. What bears the semantic weight in Representation-As are the features exemplified by X itself. In a more precise idiom we would say that X represents T as having certain features $Z_1, \ldots, Z_n$, and these features are instantiated by X. But this does not render Z (in the sense of Z-representation) otiose. As we have seen above, in order to establish that X exemplifies $Z_1, \ldots, Z_n$, one has to turn to features outside of X itself and this is where Z is

7.5 Defining Representation-As

crucial. When X is a boxer-picture (and therefore Z = boxer) the features that are exemplified are those that we typically associate with boxers. And it is these features that are imputed to someone represented as a boxer. If, however, the same drawing is classified not as a boxer-picture but as a Peter-Buckley-picture, then it would make features like being a loser salient (Buckley holds the record of having lost the largest number of recorded fights of any boxer), and these would be imputed to the target. If we only say "X exemplifies $Z_1, \ldots, Z_n$" then Z (in the sense of Z-representation) is eliminable. If, however, we want to also explain why X exemplifies $Z_1, \ldots, Z_n$ rather than a different set of features $Z'_1, \ldots, Z'_m$, then we can appeal to the fact that X is a Z-representation, and $Z_1, \ldots, Z_n$ are the features that are associated with Z. In this way, talk of Z-representations is an ellipse for conveying effectively which features of X are salient, and therefore exemplified by X.

As before, it's worth asking how the account deals with the remaining problems and conditions from Chap. 1 before turning to critical discussion. As with DDI-ER, Representation-As remains silent with respect to the Problem of Style. Different styles could be accommodated by grouping together different Z-representations, or in terms of the kinds of features they exemplify. Representation-As also remains silent about the Problem of Carriers. The account fares well on the conditions. The account straightforwardly meets the Surrogative Reasoning Condition through its posit that users first determine which features of the carrier are exemplified and then imputes these to the target. This act of imputation supplies a hypothesis about the target system.[14] Since this is a *hypothesis*, which, as such, can be false, the account at once also meets the Misrepresentation Condition: X can represent T as having $Z_1, \ldots, Z_n$ even if it turns out that T does not have these properties. This also offers a response to the Problem of Accuracy: an epistemic representation is accurate to the extent that the target has the features imputed, but that a representation be accurate is not build into Representation-As.[15] Like in the case of DDI-ER, the appeal to denotation ensures that it meets the Directionality Condition. Targetless representations are accommodated in an elegant way. A carrier can be a Z-representation without being a representation-of anything. Moreover, such a Z-representation can still exemplify features $Z_1, \ldots, Z_n$ in their own right, and can thus be used to gain epistemic access to those features independently of any target system. The account offers, however, no significant discussion about how mathematics enters epistemic representation and so it remains unclear how it meets the Applicability of Mathematics Condition.[16]

---

[14] Notice that the account has the resources to allow for more fine-grained hypotheses by appealing to the idea that there are multiple denotation relations between parts of the carrier and parts of the target, and to the notion that parts of the carrier may exemplify features that are not instantiated by the rest of the carrier.

[15] Here we are putting aside a highly significant issue. Elgin (2004, 2017) argues that many (if not most) epistemic representations, including those used in science, are, strictly speaking, inaccurate. But they are still of positive epistemic value. For this reason she develops a "non-factive" account of understanding. For want of space we cannot discuss this aspect of her later work here, see our (2019b), Lawler's (2019), and Le Bihan's (2019) for further discussion.

[16] Readers familiar with GE's project may notice that we have suppressed a discussion of the notion of 'symbol systems', frameworks in which particular instances of representation-as are embedded.

7.6 Unfinished Business

Representation-As has a lot to recommend it, but it's not without problems. One of the more significant issues facing Representation-As concerns its response to the Problem of Carriers. Recall the discussion of exemplification above: an object exemplifies a feature if it at once instantiates that feature and refers to it. The problem with relying on exemplification in a general account of epistemic representation is that, at least as stated, this requires that a carrier *instantiate* each feature that is imputed to the target system. This is problematic even where the carrier is a physical object. For example, previously we talked about a painting representing the Duchess as wise. Representation-As would require that the painting exemplify wisdom, which, in turn, would require that the painting instantiate it. And clearly no painting is wise.

In such cases GE employ the notion of metaphorical exemplification, a notion that requires metaphorical instantiation (Goodman 1976, pp. 50–51). "A painting that literally exemplifies 'dark' may metaphorically exemplify 'disturbing'" (Elgin 1983, p. 81). Likewise, the caricature that represents the politician as draconian metaphorically instantiates draconianism, and the statue that represents the dancer as graceful metaphorically instantiates grace. Whilst this initially seems relatively intuitive, what it means for an object to metaphorically instantiate a feature remains somewhat unclear. The primary issue relevant here is that GE's use of metaphor is intimately tied up with their nominalism (this is most explicit in Goodman's work, so we'll restrict our focus to him here). A nominalist of Goodman's stripe refrains from admitting entities like properties or features into their ontological commitments. As such, talking about an object, for example a telephone box, instantiating a property, for example redness, is supposed to be elliptical for claiming that the object is such that the *predicate* corresponding to the property applies to the object: the predicate "red" (and any predicate coextensive with "red") applies to the telephone box. With this in the background, Goodman (1976, pp. 68–71) can analyse metaphorical instantiation in terms of the non-standard application of a predicate to an object:

> Metaphorical application of a label to an object defies an explicit or tacit prior denial of that label to that object. Where there is metaphor, there is conflict: the picture is sad rather than gay even thought it is insentient and hence neither sad nor gay. Application of a term is metaphorical only if to some extent contra-indicated. (1976, p. 69)

Now, irrespective of whether this strategy succeeds in the context of nominalism, it is difficult to see how it would work for someone who takes discussions of properties and features at face value. For them, to say that an object instantiates a property metaphorically at once seems to say that the object has that property (because it instantiates it), and that it doesn't have it (because the instantiation is metaphorical).

To the best of our knowledge, this aspect of their project has not been explicitly discussed in the context of scientific representation, beyond a brief discussion in Gelfert's (2015) who emphasizes the communal nature of such systems. This is an area that is worthy of further research.

7.6 Unfinished Business

So what is needed is an account of metaphorical instantiation that can be used by the nominalist and the realist at the same time. To pre-empt a discussion to come, this account should also pay heed to the role played by the features an object literally has in establishing the ones that they "have metaphorically". When we say that a picture is "sad" or "gay" we do so with an appreciation of the actual features of the picture (i.e. its colour and shape).[17]

The issue with instantiation is even more pressing in the context of scientific modelling where, as we have seen in Sect. 3.4, carriers often aren't the sort of objects that instantiate, at least in any straightforward sense, physical features. This causes problems for an account of representation based on property, or feature, sharing because such carriers do not instantiate the kind of physical features that target systems have, or are represented as having. The same problem arises in the current context. If exemplification is possession plus reference, then only something that can itself move in an elliptical orbit can represent a planet as moving in an elliptical orbit. There is at least a question whether something in the Newtonian model can do this. Elgin is well aware of this problem:

> Since exemplification requires instantiation, if the model is to represent the pendulum as having a certain mass, the model must have that mass. But, not being a material object, the model has no mass. It cannot exemplify the mass of the pendulum. Indeed, the model does not exemplify mass. Rather, it exemplifies an abstract mathematical property, the magnitude of the pendulum's mass. Where models are abstract, they exemplify abstract patterns, properties, and/or relations that may be instantiated in physical target systems. It does no harm to say that they exemplify physical magnitudes. But this is to speak loosely. Strictly speaking, they exemplify mathematical (or other abstract) properties that can be instantiated physically. (2017, p. 258)

The question then, is whether either of these approaches, metaphorical exemplification or the appeal to mathematical features that can be instantiated physically, are sufficiently general to accommodate all cases of epistemic representation. As we have seen in Chaps. 3, 4 and 6, it's not clear that all non-material models should be understood as being nothing over and above mathematical objects, and it's even less clear whether all representation should be reduced to mathematical representation. But if models are understood as having features other than mathematical features, and as representing their targets as having non-mathematical features, then it is unclear how an approach based on mathematical features can account for such cases. If metaphorical exemplification is the way to deal with these models (as well as material models which we might speak of as having features they don't literally have), then can we say anything more about what it means for a carrier to metaphorically exemplify a feature and how this connects to the features the carrier literally has? As it stands, this aspect of Representation-As needs further development if it is to serve as a full-fledged account of epistemic representation.

A second problem facing the account was briefly mentioned at the end of Sect. 7.3. In order for the notion of a Z-representation to carry the theoretical weight required of it – namely to be able to account for targetless models and draw a

---

[17] For more on the role of metaphor in exemplification see Young's (2001, pp. 73–76).

connection with the features of a carrier – we need a general account of what it means for a carrier to be classified as a Z-representation. As mentioned previously, GE solve this issue by appeal to genres. Griffin-pictures, on their view, are representations because they are animal-pictures, and some animal pictures denote animals. They are griffin-pictures rather than, say, crocodile-pictures because griffin-pictures constitute a genre that contains all and only the representations whose a common ostensible subject matter are griffins. Spectators have learned to classify pictures as belonging to certain genres, and being able to do so is what it means to recognise something as a so-and-so-picture.

How pictures represent is not our primary concern here. Our problem is how scientific models, and epistemic representations more generally, work, and a theory of pictorial representation does not carry over, at least in any straightforward manner, to scientific models. The Schelling model represents social segregation with a checkerboard; billiard balls are used to represent molecules; the Phillips–Newlyn model uses a system of pipes and reservoirs to represent the flow of money through an economy; the worm *Caenorhabditis elegans* is used as model of cell division. But checkerboards, billiard balls, water pipes, and worms don't belong to classes of objects that typically denote. Most worms are just organisms and most checkerboards are just checked structures, and they don't belong to genres whose common ostensible subject matters are their respective target systems. And neither do scientific fictions such as elastically colliding point particles, frictionless planes, utility-maximising agents, and chains of perfectly elastic springs. The same problem arises with the mathematical objects used in science. Matrices, tensors, curvilinear geometries, Hilbert spaces, and Lebesgue integrals have been studied as purely mathematical objects long before they became important in the sciences. Their representational use is neither grounded in their membership in a class of representational objects (there are also objects such as quaternion groups which are similar to the ones just mentioned but which are not used representationally); nor do they belong to genres. In fact, they can represent very different things in different contexts. A matrix, for instance, can represent transition probabilities between energy levels in an atom, payoffs in game, or the stress across a surface of a material.

An alternative account of pictorial representation appeals to spectators' perceptual experience when seeing a picture rather than to genres: what kind of picture a picture is depends on our perceptual experience when looking at the picture. A picture is a unicorn-representation because we experience seeing a unicorn when looking at the picture. Different accounts unpack the notion of a perceptual experience in different ways. Schier appeals to what he calls the "natural generativity of pictures" (1986, p. 1); Wollheim speaks of the perceptual skill of "seeing-in" when a spectator sees a certain figure in a painting (1987, p. 21); and Gombrich analyses pictorial representation as being based on the illusion of thinking that we are viewing something when seeing the representation (1961).[18]

---

[18] See Kulvicki's (2006b) and Lopes' (2004) for an overview of the variety of proposed accounts of pictorial representation.

7.6 Unfinished Business

Perceptual accounts face the same difficulties as GE's genre account: regardless of how plausible they may be in the context of pictorial representation, they don't seem to work in the context of scientific representation. Models are not classified as Z-representations because they show or portray a Z. There is no "natural generativity", or "seeing-in", associated with models, and we have no illusion of viewing a gas when working with the billiard ball model.

So neither genres nor perceptual experience help us understand what makes the objects of scientific modelling Z-representations. This does not show that the conditions in Representation-As are wrong; it shows that the notion of a Z-representation needs to be explicated in further detail.

A final issue facing Representation-As concerns the idea that the exemplified features of a carrier are imputed directly to its target.[19] Whilst some examples of representation-as are ones where a feature exemplified by X is identical to the feature imputed to the target, it's far from obvious that this holds in general in pictorial representation, let alone in scientific representation, or in epistemic representation more generally. In the quote in which Elgin defines representation-as she says that X exemplifies certain properties and imputes "those properties or related ones" to the target (Sect. 7.5). This is an important qualification, as Elgin herself points out:

> In being a [Z]-representation, [X] exemplifies certain properties and imputes those properties or related ones to [T]. 'Or related ones' is crucial. A caricature that exaggerates the size of its subject's nose need not impute an enormous nose to its subject. By exemplifying the size of the nose, it focuses attention, thereby orienting its audience to the way the subject's nose dominates his face or, through a chain of reference, the way his nosiness dominates his character. (Elgin 2017, p. 260)

In this example the caricature exemplifies the feature of having a large nose. But it's explicitly not this feature that is imputed to the target. Rather the feature is connected, via a "chain of reference", to a "related one", nosiness as a character trait, and it's this trait, not the feature literally exemplified by the caricature, which is imputed to the target.

It's an interesting question as to how committed Elgin is to this clause. Despite emphasising the importance of "related ones", little use is made of the clause in her analysis of scientific representation and she also writes:

> for [X] to exemplify a property of [T], [X] must share that property with [T]. So [X] and [T] must be alike in respect of that property. It might seem, then, that resemblance in particular respects is what is required to connect a representation with its referent [...]. There is a grain of truth here. If exemplification is the vehicle for representation-as, the representation and its object resemble one another in respect of the exemplified properties. (2017, p. 261–62)[20]

---

[19] This issue is made more complicated in the light of the previous discussion of metaphorical exemplification: if one grants that carriers can metaphorically exemplify features that they do not strictly speaking instantiate, then the question is whether or not the metaphorically exemplified features are imputed directly to the target. As we will see in the next chapter, unpacking these issues is required for a systematic account of epistemic representation.

[20] van Fraassen (2008, p. 17ff) also talks about exemplification as highlighting the relevant similarities between carrier and target, which is in line with his broadly structural approach to epistemic representation and helps explain why he also invokes "representation-as".

Moreover, as we note in our (2019b) the idea that scientific models represent their targets as having features that they themselves exemplify is an important premise in her argument for non-factive understanding. So it remains unclear what her considered position on the matter is.

Regardless, the clause is required in order to make sense of many cases of epistemic representation. We have already seen the example of the caricature with the large nose. The size of the nose is exemplified in the image, but this feature is not imputed to the target. Rather, the exemplified feature is connected to another feature – the character trait of being nosy – and it is the latter feature that is imputed to the target. It's not only caricatures that function in this way. Consider an icon by Theophanes the Cretan, a leading icon painter of the Cretan School in the early sixteenth century. The icon shows Jesus in a robe that is half red and half blue. But the icon does not represent Jesus as wearing a bi-coloured dress, and it does not impute to him having worn such attire. The icon does not function like a photograph. Icons use a colour code whereby blue, the colour of heaven, stands for the Kindom of God and another everlasting world, while red, the colour of blood, stands for life on earth and humanity. By showing Jesus wearing a red and blue dress, the icon represents him as belonging to both the realms of heaven and earth, as being a creature that is both divine and human.

Moving beyond pictorial representation, consider a map of the world with a 1 mm: 10 km scale. The map exemplifies a distance of 75 cm between the two points labelled "Vancouver" and "London". But the map doesn't represent the cities as being 75 cm apart, despite exemplifying this feature. Rather, the feature "being 75 cm apart" is connected, via the map's scale, to another feature, "being 7500 km apart", and it's the latter feature that is imputed to the cities themselves. Or consider a piece of litmus paper dipped into an acidic solution. The paper exemplifies redness, but clearly it doesn't represent the solution as being red; it represents it as being acidic. A scale model of a ship in a water tank exemplifies various features concerning the resistance it faces when towed through the tank. But these features are not carried over directly to the actual ship. In fact, in this case, they are not even simply scaled: there are complex non-linear relationships between size and viscosity, and researching how to connect the features exemplified by the model ship with the features of the actual ship is a complex part of scientific practice (we come back to these cases in Sect. 8.4). Hence, many cases of epistemic representation can only be understood if we provide explicit specification of the exemplified features of the carrier and the "related" features that are imputed to their target.

So, given that the clause is indispensable, how are we to understand the qualification "or related ones"? The problem with invoking "related" features is not its correctness, but its lack of specificity. Any features can be related to any other feature in some way or other and as long as nothing is said about what this way is, it remains unclear what features X ascribes to T. Elgin does provide a few clues about the relation between the features exemplified and the ones ascribed to the system, which

7.6 Unfinished Business

she sometimes describes as one of simplification or idealisation (1996, p. 184), or of approximation (2010, p. 11). While this is certainly the case in some instances, it's not a general account because not all representations are idealisations or approximations, at least in any obvious way. Indeed neither the map nor the litmus paper fit this mould.

So, to sum up. Whilst Representation-As is on the right track, it needs further development if it is going to serve as a fully general account of epistemic representation. First, it needs to accommodate the common phenomena of carriers exemplifying features that they don't strictly speaking instantiate. Second, the idea that epistemic representations other than pictures can be classified as Z-representations requires further explication. Finally, the notion that the features imputed to the target are often not the ones that are exemplified, but are "related" features, needs to be further developed. That is the task of the next chapter.

Chapter 8
The DEKI Account

We now address the issues faced by Representation-As (as discussed in Sect. 7.6) and develop it into a full-fledged account of epistemic representation. The result of this endeavour is what we call the DEKI account, where the acronym is formed from the names of the account's defining features: denotation, exemplification, keying-up, and imputation. To keep the discussion manageable we develop the DEKI account with scientific models. We occasionally hint at how it would deal with other kinds of epistemic representations, and we return to the issue in more detail in Sects. 9.4 and 9.5. In this chapter we introduce the DEKI account in the context of material models and discuss the account's basic tenets. In the next chapter we extend the account to non-material models.

We begin by introducing one of our favourite material models, the so-called Phillips–Newlyn machine.[1] This machine is our "conceptual laboratory" that we use to introduce the DEKI account and guide us through the discussion (Sect. 8.1). We then give an account of what makes something a Z-representation, and on the basis of that account offer a definition of a model (Sect. 8.2). Based on this account of Z-representations we revisit the notion of exemplification and introduce the concept of exemplification under an interpretation (Sect. 8.3). Few, if any, models instantiate features that are imputed to the target unaltered.[2] Exemplified features undergo a transformation with a key before being imputed to the target (Sect. 8.4). Gathering these elements results in a complete statement of the DEKI account (Sect. 8.5). The account bears on a number of other issues, in particular on scientific realism and

---

[1] The machine is often referred to as the "Phillips Machine" (see, for instance, Vines 2000). We follow Morgan (2012, Chap. 5) in referring to it as the "Phillips-Newlyn Machine" to acknowledge the contribution of Walter Newlyn to the development and construction of the machine.

[2] Recall that we are using the term "feature" in line with Elgin's use to refer to any property, relation, or pattern. There are no restrictions on these features; they can be "static or dynamic, monadic or relational, and may be at any level of generality or abstraction" (2017, p. 185).

representational content, and we end this chapter by briefly indicating what the implications of DEKI for these issues are (Sect. 8.6).

8.1 Water Pumps, Intuition Pumps

If you go to the Science Museum on London's Exhibition Road, you can find among its exhibits a curious machine. Figure 8.1 is a photo of the machine taken on the occasion of our own visit to the museum. The machine is about 7 feet high, 5 feet wide, and 3 feet deep. It consists of transparent reservoirs that are connected by

Fig. 8.1 The Phillips–Newlyn machine in the London Science Museum with the two authors

clear Perspex pipes, which are controlled by various valves.[3] When in operation, the machine contains red water which flows through its various channels from the top of the machine to the bottom, where it enters a pump which transports it back to the top. In this way the machine creates a circular flow of water through its systems of pipes and reservoirs, which is clearly visible due to the see-through nature of the reservoirs and tubes. The specific characteristics of the flow depend on the settings of the valves.

But why, you may now ask, is this object on display in a *science* museum? Per se the machine is just a curious piece of unorthodox plumbing and one might think that it is oddly out of place in a museum dedicated to the history of learned enquiry. This impression is mistaken: what you have in front of you is an economic model! The model was initially developed by Bill Phillips and Walter Newlyn in the late 1940s; a prototype was available in 1950; and north London engineering firm White-Ellerton Ltd. produced them throughout the 1950s. The core idea of the machine is to translate economic concepts into hydraulic ones by using water to represent money. Water in reservoirs can then be seen as stocks of money and water running through plastic tubes as monetary flows, and in this way each reservoir and each pipe in the machine comes to represent a different element of an economy.

We now give an outline of how the machine works. Our outline closely follows Barr's presentation in his (2000, pp. 100–04). We keep this short for two reasons. First, as noted in the introduction to this chapter, in the current context the machine serves as a "conceptual laboratory": it helps us develop, test, and illustrate concepts in our account of representation. To this end it is not only unnecessary to go deep into the economic thinking behind the machine; doing so would in fact be a distraction. Second, as several commentators have pointed out, the machine is essentially visual: when seeing the machine it is clear how it works, but written accounts inevitably become laborious and unenlightening.

Figure 8.2 shows an extremely simplified version of the machine, which explains the basic structure of the circular flow through the machine. At the bottom there is the reservoir with transaction balances. These balances are pumped up to the top of the machine through the tube on the left, and when at the top they enter the economy as income. Once at the top of the machine, gravity ensures that the water will work its way down. Income divides into three parts. On the left there is the government sector, which collects taxes; in the middle there is the consumption flow; and on the right savings leave the circular flow to go into a reservoir labelled "idle balances", which can later re-enter the flow as investments. Below the savings is the foreign trade sector, consisting of imports leaving the domestic spending cycle and exports

---

[3] The Science Museum provides some basic information about the machine on its website at https://www.sciencemuseum.org.uk/objects-and-stories/how-does-economy-work. The machine can be seen in action in various online videos; one of our favourites is https://www.youtube.com/watch?v=k_-uGHWz_k0&t=1s. Detailed discussions of the machine can be found in Barr's (2000), Morgan's (2012, Chap. 5), Morgan and Boumans' (2004), Newlyn's (1950), Phillips' (1950), and Vines' (2000).

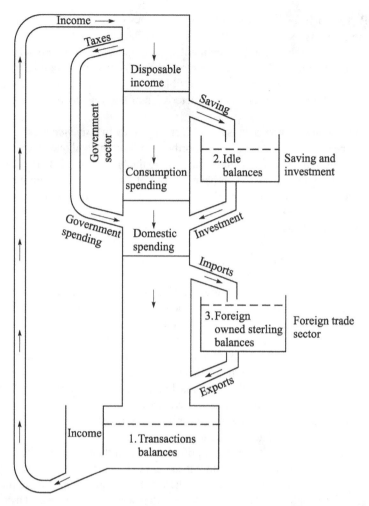

Fig. 8.2 Extremely simplified version of the Phillips–Newlyn machine. (Barr 2000, p. 101, reprinted with permission of the author)

coming back into it. At the bottom; the domestic spending enters the reservoir as transaction balances and thus completes the cycle.

The equilibrium condition of the simple economy in the machine can be formulated as follows. Total income T enters the machine at the top. Taxes are deducted from T, leaving disposable income. The disposable income divides into consumption C and savings, which are funnelled back into the flow as investments I. Taxes return to the central column as government spending G. The total domestic spending therefore is $C + G + I$, that is the sum of consumption, investment, and government spending. From this, imports J are deducted, and exports E are added. This gives the equilibrium condition for the model-economy, which is also the equation that governs the circular flow of water through the machine:

8.1 Water Pumps, Intuition Pumps

$$T = C + G + I + E - J.$$

The most significant simplification in the version of the machine shown in Fig. 8.2 is that the machine has neither controls nor feedbacks. This simplification is dropped, at least to some extent, in the more detailed presentation of the machine in Fig. 8.3, where a system of valves controls the inflows and outflows in various parts of the machine. Some of the valves are controlled from the outside. For instance, how much of the income flow goes into tax and how much into disposable income is determined by a slider that shows a functional relationship between tax and income, and varying the position of the slider determines how much income makes its way through the taxation pipe. Other valves respond to what happens in other parts of the machine, for instance by connecting a float on top of a reservoir to a valve via a cord that exerts a downward pull on a valve when the water level in the reservoir falls. Thus, when income goes down, the consumption valve will partly close and thereby reduce consumption. Finally, the machine has a number of recorders. These are clockwork mechanisms with a chart that move a pen horizontally along rails at a rate of one inch per minute while recording the magnitude of a certain quantity vertically. In this way the machine draws a graph of the quantity against time.

Theoretically speaking the Phillips–Newlyn machine is a so-called open economy IS-LM model. It is "open" because it includes international trade; "IS" stands for "investment-savings"; and "LM" for "liquidity preference-money supply" (Barr 2000, p. 103).[4] The theoretical setup of the machine is extremely flexible, allowing the operator to choose settings that correspond to either a "Keynesian" economy (in which expansionary fiscal policy is effective in increasing output and employment), or a "classical" economy (in which fiscal policy has no effect on output). The machine can run various scenarios and produce results that are ±4% accurate with respect to the underlying theoretical model.[5]

Throughout the 1950s, the machine was used in university classrooms to teach students macroeconomics. However, the machine's use was not limited to pedagogical applications, and it was used as a tool to develop economic analysis.[6] A number of experiments have been made with the machine and various conclusions have been drawn.[7] We here only briefly mention two. The first result is that the machine resolved a controversy between Keynes and Robertson over monetary theory. According to Keynes, the equilibrium interest rate is determined by liquidity preference; according to Robertson it is determined by the supply and demand of

---

[4] For a discussion of this model see Begg et al. (2014, Chap. 20).

[5] The machine can therefore also be regarded as calculator that determines the values of variables in a theoretical model. This explains why it is also known as the MONIAC (Monetary National Income Analogue Computer) and is on display in the computing section of the London Science Museum. That it is an analogue computer does nothing to detract from the fact that it also serves to function representationally.

[6] Vines calls the Phillips–Newlyn machine a "progressive" model to emphasise this aspect (2000).

[7] For a discussion see the contributions in Part II of Leeson's (2000).

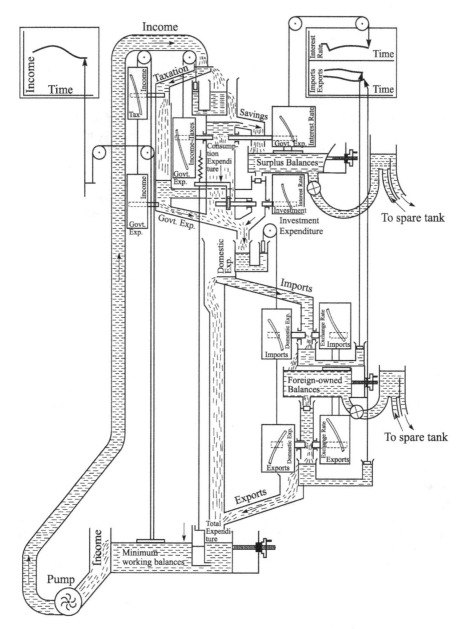

Fig. 8.3 More detailed version of the Phillips–Newlyn machine. (Barr 2000, p. 103, reprinted with permission of the author)

loanable funds. The machine showed that, and how, in equilibrium both points of view are correct (Barr 2000, p. 108).

The second concerns stability. The machine shows that control, if it is applied strongly and with a lag, can introduce a cycle into the system. This shows that a badly thought out policy response that is brought in with a lag can destabilise an economy (Vines 2000, p. 53). A related result emerged in James Meade's seminar. He had convinced the LSE to buy a second machine and coupled it to the first, so that exports of one machine became the imports of the other. The two machines together formed a two-country world economy, which made it possible to investigate the effect of, say, a budget deficit in one country on the economy of the other country. Meade used this setup to show the destabilising consequences of ill-considered policy interventions. He did this by making a student Chancellor of the Exchequer, tasked to manipulate taxes and government spending with the aim of achieving a target level of national income, and another student Governor of the Bank of England, tasked with making monetary policy to the same end. With the two machines connected, he then added the US Secretary of the Treasury and the Chairman of the Federal Reserve. The four students had complete freedom in their choices, the only instruction being that each student act independently and ignore the others. The results demonstrated impressively the destabilising consequences of ill-considered policy interventions, and how quickly a destabilising policy in one country can be transmitted to another country in an inter-linked economy (Barr 2000, pp. 108–09).

Returning to the notions we introduced in Sect. 7.3, we can say that the Phillips–Newlyn machine is, by construction, an economy-representation, or, more specifically, an open-IS-LM-economy-representation. This status requires no target system, and the machine would be an economy-representation even if had never been used to represent an actual economy in the world. Turning the machine into a representation-of an economy requires an additional step. In Meade's seminar one of the machines was used as representation-of the UK economy and the other as a representation-of the US economy. A similar thing happened in Guatemala in 1953. The land reform act passed in Guatemala the previous year had redistributed unused land to local farmers. US corporation Wrigley's, one of the largest buyers of Guatemalan chicle gum, had announced that it would stop imports from Guatemala in protest to the land reform. The economists in the Central Bank wanted to know what effect a decrease in these foreign purchases would have on the national economy. They turned to their Phillips–Newlyn machine in the hope that they would get useful answers.[8] Clearly, the Guatemalan central bankers used the machine as a representation-of the Guatemalan economy.

---

[8] There are records showing that the Guatemalan central Bank purchased a machine (Aldana 2011; Stevenson 2011). However, central banks are notoriously secretive about their activities and there are no (publicly available) records of the exact uses the machine has been put to, so our example is illustrative rather than a statement of historical fact.

Now you wonder: how is it that a system of pipes and reservoirs is an economy-representation and how is such an economy-representation subsequently turned into representation-of a particular economy such as the UK's or Guatemala's?

Before addressing these questions we would like point out that the Phillips–Newlyn machine is not a curiosity, judiciously chosen to serve our purposes. We chose the machine as our example of a material model because of its intuitive appeal and because (at least the broad outlines of) its method of operation can be understood without much theoretical background. Nevertheless, the Phillips–Newlyn machine is just an example and we could make our points equally well with other examples. Plasticine sausages are used as models of proteins; oval shaped blocks of wood serve as models of ships; small-scale reproductions of cars serve as models of real cars; mice are used as models for humans; balls connected by sticks function as models for molecules; electrical circuits are studied as models of brain function; autonomous robots are used as models for insect cognition; a *camera obscura* is proffered as a model of the human eye; metal cylinders filled with hardened magma are investigated as models of volcanoes; a basin with pumps and hoses serves as a model of the San Francisco Bay's water system; and fluid "dumb holes" are used as models of gravitational black holes.[9] In all these cases, and indeed in many others, a material object is used as a model that represents a certain target system. Our question about the Phillips–Newlyn machine is therefore an instance of a general problem: when is a material object a Z-representation and how does a Z-representation become representation-of a target? These are the questions that we address in this chapter, and their answers will lead to the DEKI account of epistemic representation.

8.2 Models and Z-Representations

How do we categorise objects as Z-representations? As we have seen in Sect. 7.6, an appeal to spectators' visual experience is a non-starter, and reference to genres at the very least requires unpacking. Reservoirs and pipes, plasticine sausages, blocks of wood, mice, balls and sticks, electric circuits, robots, the camera obscura, and metal cylinders are not in any obvious way classified as belonging to a particular genre of representations, and most objects of this kind don't usually function symbolically at all. In the context of material models an alternative approach to understanding Z-representation is needed.

---

[9] For discussions of proteins see de Chadarevian's (2004); for ships see Sterrett's (2002, 2006) and Leggett's (2013); for cars see Cyr et al. (2011); for model organisms see Ankeny and Leonelli's (2011); note, however, that classification of organisms as models has recently been opposed by Levy and Currie (2015). For molecules see Toon's (2011); for brain functions see Sterratt et al. (2011); for insects see Webb's (2001); for the *camera obscura* see Wade and Finger's (2001); for volcanoes see Spieler et al. (2004); for the San Francisco Bay model see Weisberg's (2013); and for "dumb holes" see Dardashti et al. (2017).

8.2 Models and Z-Representations

The carrier of the Phillips–Newlyn model is the system of reservoirs and pipes that Phillips and Newlyn built, the physical object that one can now see in the London Science Museum. Likewise, the carrier of the protein model is the plasticine sausage; the carrier of the molecular model is a particular configuration of balls and sticks; the carrier of the San Francisco bay model is the water basin with pumps; and so on. We speak of *X-features* when we want to emphasise that we are referring to the features that the carrier has *qua physical object* and not qua representation. The Phillips–Newlyn machine consists of valves and pipes and reservoirs, and its X-features are features like a reservoir containing so and so many litres of water, being approximately 2 m high, there being a leak in the bottom pipe, and so. There is no money in the carrier *qua physical object*, and features like consumption being at a certain level or embodying the principles of Keynes' economic theory are not X-features. To emphasise that features pertaining to something *qua* Z-representation we speak of *Z-features*. In the context of the Phillips–Newlyn machine the Z-features include features like there being such an interchange between different areas of an economy.

The question then is: what turns X into a Z-representation? An appeal to X's intrinsic features does not help. There is nothing in water pipes or electric circuits that makes them economy-representations or brain-representations, and the mouse running through the kitchen isn't a representation at all. X's intrinsic characteristics do not regulate how the object functions representationally. One might say that someone using X as such is what turns it into a Z-representation. There is a grain of truth in this, but it merely pushes the question one step back: what does it take to use an X as a Z-representation?

As we have seen in the last section, the idea behind the machine is that hydraulic concepts are made to correspond to economic concepts. This is means that we turn system of pipes and reservoirs into an economy-representation by *interpreting* certain selected X-features as Z-features. The water in a certain reservoir is interpreted as money being saved; the level of water in the reservoir is interpreted as a quantity of money; and so on.

"Interpretation" is a flexible term that can mean different things in different contexts. It is therefore important to define we mean by interpretation. The leading idea here is that an interpretation connects features of the carrier to features in the domain of Z. This can be formalised as follows. Consider a set $\mathcal{X} = \{X_1, ..., X_n\}$, where the X_i are X-features, and a set $\mathcal{Z} = \{Z_1, ..., Z_n\}$, where the Z_i are Z-features, and n is an integer. A first stab at defining an interpretation then is to say that it is a bijective function $I : \mathcal{X} \to \mathcal{Z}$. While correct in principle, this definition is not easy to handle in practice because it does not explicitly distinguish between quantitative and qualitative features.[10] Features like "being a valve" or "being a reservoir" are *qualitative* features: they are all-or-nothing in that they either are or aren't instantiated. By contrast, features like "the flow of water is y litres per minute" are

---

[10] See Eddon's (2013) for a detailed discussion of quantitative properties (sometimes called "quantities").

quantitative features. In that case we need to distinguish between the property and its values. To make this distinction explicit we refer to the feature itself as the *variable* and to a specific quantity as the *value*. We denote the former by upper-case letters and the latter by lower-case letters. So X_1 could be the flow of water through the second valve, and x_1 could be 2.3l. This suggests the following definition of an interpretation:

Interpretation: Let $\mathcal{X} = \{X_1, ..., X_n\}$ be a set of X-features and $\mathcal{Z} = \{Z_1, ..., Z_n\}$ a set of Z-features. The elements of both sets are either qualitative features or variables. An interpretation is a bijection $I : \mathcal{X} \to \mathcal{Z}$ so that:

(i) each X-feature is mapped onto a Z-feature of the same kind (that is, a qualitative property is mapped onto a qualitative property and a variable onto a variable);
(ii) for every variable X_i, mapped onto variable $Z_j = I(X_i)$ under I, there is a function $f_i : o_i \to z_j$ associating a value of Z_j with every value of X_i.

As an example consider $\mathcal{X} = \{X_1, X_2\}$ with $X_1 = $ "quantity of water" and $X_2 = $ "time elapsed in the machine", and $\mathcal{Z} = \{Z_1, Z_2\}$ with $Z_1 = $ "quantity of money" and $Z_2 = $ "time elapsed in the economy". The interpretation we have discussed so far maps X_1 onto Z_1 with the scaling function $f(x_1) = 10^6 x_1$ (i.e. f specifies that that the quantity of money is a linear function of the quantity of water with 1 l of water corresponding to a million of the model currency). Processes both in the machine and in an economy take time to unfold. The settings of the Phillips–Newlyn machine can be chosen such that they introduce a time-scale, making, for instance, 1 min of running time in the machine equivalent to 1 year in the economy (Phillips 1950, p. 286). The interpretation therefore also maps X_2 onto to Z_2 with the function $f(x_2) = 525600x_2$ (because a calendar year is equivalent to 525,600 min). Sometimes one may want to impose further restrictions on the allowable functions. In the case of the Phillips–Newlyn machine the functions are assumed to be linear. However, such restrictions are specific to the context and should not be built into a general definition.

We are now in position to define a Z-representation:

Z-representation: A Z-representation is a pair $\langle X, I \rangle$ where X is an object and I is an interpretation.

Colloquially one can call an X a Z-representation, and as long as it is understood that there is an interpretation in the background no harm is done. It is important, however, that in a final analysis a Z-representation is a pair $\langle X, I \rangle$ and an object X is never a Z-representation just on its own accord: no interpretation, no Z-representation! The system of pipes you see in the Science Museum is an economy-representation only under (something like) Phillips and Newlyn's interpretation that takes water to be money. Without such an interpretation the machine is just an eccentric piece of plumbing. What kind of representation X is crucially depends on I, and different interpretations produce different representations. One could, for instance, take the Phillips–Newlyn machine and interpret the reservoirs as schools and the flow of water as the movement of pupils between different schools. Under that interpretation the same machine would be an education-system-representation.

8.2 Models and Z-Representations

Hence, what turns an X into a Z-representation is an interpretation. In some cases X is carefully constructed to be interpretable in a certain way. Phillips and Newlyn carefully designed and constructed their reservoir and pipe system so that it is interpretable as an economy. The same is true of the San Francisco Bay model, autonomous robots, models ships, ball and stick models, and hydrodynamic dumb holes. But not all carriers are tailor-made. Some are, as it were, readymades. Model organisms like the worm *Caenorhabditis elegans* and the plant *Arabidopsis thaliana*, and physical contraptions like the electric circuit predate their use as models.[11] The choice of a carrier is a creative act. It may be informed by the interpretation that one would like to impose on an object, but it is in no way determined by it. But irrespective of whether objects are purposefully constructed or serendipitously found, what converts them from a "mere" object into a Z-representation is subjecting them to an interpretation. This is why, in principle, anything can serve as Z-representation.[12]

We are now in a position to say what a model is:[13]

Model: A model M is a Z-representation, $M = \langle X, I \rangle$, where X is chosen by a scientist (or a group of scientists) to be a carrier in a certain modelling context and I is an interpretation.

An immediate consequence of this definition is that models need not have a target. Far from being an unintended flaw, this is an advantage of our account. It provides a natural answer to how models without targets represent: they are Z-representations that are not also representations-of a target. The Phillips–Newlyn machine would be an economy-representation even if it had never been used as a representation of an actual economy (UK, USA, Guatemalan, or otherwise), just as an architectural model of Gaudi's Hotel Attraction is a Hotel-Attraction-representation even though the hotel was not built. If the Guatemalan central bankers hadn't used their machine to be a representation-of the Guatemalan economy, the machine would never have been a representation-of the Guatemalan economy; and if no one had ever used the machine to represent any particular economy, then the machine wouldn't be a representation-of anything. It would, however, still have been an economy-representation, which could be used to study the relations between certain economic concepts.[14] For some applications nothing beyond being a Z-representation is needed. Phillip's resolution of the controversy between Keynes and Robertson didn't rely on the

---

[11] For a discussion of model organisms see, for instance, Ankeny and Leonelli's (2011) and Leonelli's (2016).

[12] This observation is closely related to the often-made point that anything can be a model of anything else (Callender and Cohen 2006, p. 73; Frigg 2010b, p. 99; Giere 2010, p. 269; Suárez 2004, p. 773; Swoyer 1991, p. 452; Teller 2001a, p. 397).

[13] This idea is similar to Danto's characterisation of an artwork as an interpreted object; see Frigg's (2013).

[14] In fact, Phillips and Newlyn's own motivations for building the machine were not to represent any particular economy. Rather charmingly, Phillips describes building the machine in order to help "students of economics who, like [himself], are not expert mathematicians" (1950, p. 283) understand the mathematical equations that were increasingly being used in macroeconomics at the time.

machine being representation-of the UK economy, or, indeed, any other economy; it only relied on the machine being an open-IS-LM-economy-representation. In sum, all models are Z-representations, but not all models are representations-of a target.

Carriers can be chosen freely, and our notion of interpretation imposes no restrictions on the choice of either X-features or Z-features. Did we open the floodgates to arbitrariness? No. Modellers will choose carriers that exhibit interesting behaviours. As we noted in Sect. 7.2, Hughes rightly observes that (what we call) the carrier is a secondary object with "a life of its own" and an "internal dynamic" that we can examine (1997, p. 331). The Phillips–Newlyn machine is a case in point. The machine exhibits a highly complex behaviour that economists can study. Even though they could have built the machine differently, or they could have chosen to study another object altogether, once the choice is made, what they see is far from arbitrary. Scientists will choose as carriers objects that have interesting features and that exhibit interesting behaviour. But this choice is pragmatic. From a semantic point of view nothing stops us from taking a pencil to be a carrier and turning it into a universe-representation by saying that the lead is the Big Bang, the body is the universe after the Big Bang, and the eraser at the top is the Big Crunch. This interpretation *does* turn the pencil into a universe-representation, but this interpretation will be useless because the pencil does not exhibit behaviour that allows scientists to learn anything interesting from the model. But that does not show that the pencil is not a model; it shows that it's a useless model.

At this point one might object that models have to have an internal dynamic, but pencils, along, perhaps, with drawings, photographs, and maps, are static objects. These, however, also meet the conditions for being a model because they are objects with interpretations. So, on our account, static objects can count as models, but this, so the objection goes, is implausible. Whilst there is something to this objection, we think that it is ultimately misplaced. In order for something to be the sort of Z-representation that is useful as a model, it is often the case that the carrier is chosen such that it has a complex behaviour, thus generating a complex set of Z-features that can be investigated. This is precisely what we see with the Phillips–Newlyn machine, and this complexity is usually associated with having an internal dynamic. But both of these points are contingent: models don't have to be complex, and complexity can be found without exploiting an internal dynamic. Moreover, these considerations are more adequately thought of as concerning the *styles* of different epistemic representations, rather than a difference with respect to how epistemic representations work *qua* representations. Whilst, as a matter of fact, many objects that function as models have an internal dynamic, this shouldn't be turned into a defining feature of models, let alone Z-representations. Whether one wants models to have an internal dynamic is a pragmatic question that depends on what scientists in a certain context aim to achieve.

Similar remarks apply to interpretations. Scientists are free to choose X-features and Z-features as they please. But once a choice is made, representational content is constrained. If there are three litres of water in the reservoir and the interpretation says that the reservoir holds foreign-owned balances and 1 l of water corresponds to

8.2 Models and Z-Representations

one million of the model currency, then the model says that there are one million of the model currency outside of the country. Free choices, once made, are highly constraining. This is why models are epistemically useful. Scientists study this constrained behaviour and thereby gain insight into their subject matter.

A few observations about Z-representations are in order. First, Z can be a concept, a notion, an idea, or a phantasy – anything that can belong to a certain domain of discourse. The important point is that Z need not pertain to a target system. Indeed, Z need not be realistic at all. A drawing can be minotaur-representation, a movie can be a Darth-Vader-Representation, and a model can be a perpetual-motion-machine-representation or a superluminal-beam-travel-representation, yet none of these exist. There are no limits to the choice of Z; anything that makes sense in a certain context is in principle acceptable. This frees Z-representations from the dictate of targets.[15]

Second, our definition of a Z-representation does not require that *all* of X's features be collected in $\mathcal{X}$; neither does it require that $\mathcal{Z}$ contain a *complete* list of Z-features. All that is required is that there is at least one feature in each set and that the two features are connected through an interpretation. The sets can be changed, which leads to a different interpretation.

Third, interpretations aren't set in stone. In different contexts the features that occur in $\mathcal{X}$ and $\mathcal{Z}$ (and the details of the interpretation function itself) may change, and existing interpretations can be extended. The Phillips–Newlyn machine often leaked water onto the floor when it was run. Originally this was seen as technical problem with the machine. However, at some point economists realised that this was actually an interesting feature and interpreted it as the flow of money from the regular economy into the black economy (Morgan and Boumans 2004, p. 397 fn. 14). So "leaking water" was included into $\mathcal{X}$ and endowed with an interpretation mapping it to "money flowing into the black economy".

Fourth, nothing in the notion of an interpretation requires that $\mathcal{X}$ and $\mathcal{Z}$ be sets containing distinct features. In some cases, an object X may be interpreted in terms of X-features: features it has as a carrier. This is what happens, for instance, in scale models. The small ship is a ship-representation where X is a ship-object and the model's interpretation takes ship-features to ship-features. It's also the case for Watson and Crick's scale model of DNA, which is a helical structure that is interpreted in terms of helical-structure-features, and for the Army Corps' model of the San Francisco Bay, which is a water-current-object that is interpreted in terms of water-current-features.

---

[15] Our account differs from Contessa's (2007, p. 58), who uses the term "interpretation" to describe a one-to-one association of the relevant objects, relations, and functions, found in models and their targets. It also differs from Weisberg's (2013, pp. 39–40) notion of a "construal", which includes an "assignment" (denotation relations between a model and its target) and a specification of the "intended scope" (a specification of which aspects of the target the model is supposed to represent) to capture an idea similar to Contessa's. We discuss this in more detail in Sect. 9.1.

8.3 Exemplification Revisited

What we are steering at is an account in which a model exemplifies certain features. As we have seen in Sect. 7.4, an item exemplifies a feature P iff it instantiates and highlights P. How does a model meet these conditions?

Let us begin with instantiation. A model $M = \langle X, I \rangle$ is a composite entity consisting of a carrier X and an interpretation I. What does it mean for such an object to instantiate a feature? An obvious answer seems to be that M instantiates P iff X instantiates P. This might work in with the sorts of models discussed at the end of the last section, models whose interpretations link features of the carrier with themselves. However, it clearly fails in cases where the interpretation takes X-features to non-X-features, because in such cases P is a non-X-feature, and therefore is not instantiated by X. The Phillips–Newlyn machine is case in point. X is a pipe-and-reservoir-system and its X-features are pipe-and-reservoir-features. But Z contains economy-features and the machine plainly does not instantiate economy-features.

GE resolve a related problem – in their case that a painting does not instantiate features like wisdom – by appeal to metaphorical instantiation. However, as we have seen in Sect. 7.6, this answer is unsatisfactory for several reasons. It is tied up with their nominalism, and it fails to sufficiently highlight the important role played by the features *literally* instantiated by objects in "grounding" the features they metaphorically instantiate. Fortunately, the notion of an interpretation, which we introduced in the previous section, offers an elegant solution. An interpretation establishes a one-to-one correspondence between X-features and Z-features and so we can introduce the concept of *instantiation-under-interpretation-I* (*I-instantiation* for short):

I-instantiation: a model $M = \langle X, I \rangle$ I-instantiates a Z-feature Z_j iff X instantiates an X-feature X_i such that: X_i is mapped to Z_j under I, and if X_i and Z_j are variables then I contains a function f such that $z_j = f(x_i)$.

In brief, a model I-instantiates a certain Z-feature if the carrier instantiates the corresponding X-feature, where the correspondence between features is given by the interpretation I.

It is now just a short – and trivial – step to liberalise the definition of exemplification to allow objects that I-instantiate a feature to exemplify that feature: we simply have to replace all occurrences of "instantiation" by "I-instantiation" in the definition in Sect. 7.4. The result of this substitution is our following definition of I-exemplification:

I-Exemplification: model $M = \langle X, I \rangle$ I-exemplifies Z-feature Z_j in a context C iff M I-instantiates Z_j and Z_j is highlighted in C, whereby Z_j is *highlighted* in C iff (i) C selects Z_j as a relevant feature, and (ii) Z_j is epistemically accessible in C.

We can now say that the Phillips–Newlyn machine I-exemplifies particular economic features.

8.3 Exemplification Revisited

Some comments about a Z-feature being highlighted are in order. Let us begin with condition (ii) and get clear on what it means for Z_j to be epistemically accessible. The answer, we submit, follows directly from the notion of an interpretation: Z_j is epistemically accessible iff X_i is epistemically accessible, where X_i is the property that is mapped to Z_j under I. In this way an interpretation carries epistemic access to an X-feature over to epistemic access to a Z-feature.

Condition (i) is less straightforward. The worry here is that all features in $\mathcal{Z}$ are *automatically* selected as relevant and epistemically accessible when the set $\mathcal{Z}$ was constuctured, and that therefore all features in $\mathcal{Z}$ are always exemplified. If so, this would trivialise the notion of exemplification. This worry is based on a misapprehension about the workings of an interpretation. Interpretations can, and often do, cover X-features we are unaware of, or uninterested in. Suppose that there is a tiny pipe somewhere at the bottom of the machine, and that we haven't paid any attention to either the pipe or the flow through it. This flow is covered by the interpretation, but it is not highlighted and therefore not exemplified. Or consider again the case of the Guatemalan economists. They may have been particularly interested in the change in the equilibrium values once the appropriate change had been made to the valve marked "foreign exports". This means that the machine exemplifies features pertaining to foreign export. But in other contexts, these might not be exemplified at all. For example, when explaining the working of the machine, Phillips himself ignores the impact of foreign imports and exports until the end of his paper (1950, Sec. III), which means that, although I-instantiated, the relevant Z-features would not be highlighted, and thereby would not be exemplified even though they have been covered by the interpretation all along. Or consider again the leakage problem mentioned previously. The machine was leaking water, and the water was covered by the interpretation. So the machine instantiated the feature of the economy losing money. This, however, was not highlighted until someone saw this as relevant to understanding the black economy.

Furthermore, as we have seen in Sect. 7.4, exemplification is selective. Not every feature that is somehow salient is also exemplified. In the case of the Phillips–Newlyn machine, the model exemplifies how the water flows through the machine, and the relative height of the liquid in its various reservoirs through time. It doesn't exemplify being made of Perspex or being 7 foot tall. And it's not just features that are irrelevant to the workings of the machine that are not exemplified. In order for the water to move around the model at all, it requires a motor that pumps water from the floor level reservoir up to the top of the machine, and the force of gravity that draws the water downwards through the various pipes and reservoirs. Although these aspects of the machine are essential to its workings, they do not correspond to any economic feature (Morgan and Boumans 2004, p. 386). They are not selected as relevant features of the machine in the context of using it as an economic model, and so are not exemplified.

8.4 Imputation and Keys

So far we have focused on discussing what turns something into a Z-representation, and GE's observation that Z-representations do not have to be representations-of anything remains intact in the context of scientific representation. Yet, at least some models do represent targets. A necessary condition for this to happen is the model has to exemplify features and impute these to the target T.

Imputation can be analysed in terms of feature-ascription. The model user might simply ascribe to the target system the features exemplified by the model, and this is what establishes that the model represents the target *as* having those features. In this way models allow for surrogative reasoning: imputing a feature to a target generates the hypothesis that T has that feature. As previously noted, nothing in our discussion requires that these hypotheses are true. The result of an imputation can be right or wrong, thereby allowing models to misrepresent their targets. Indeed, the model user needn't even believe the hypothesis, thereby allowing this misrepresentation to be deliberate.

However, as noted in Sect. 7.6, in many cases of representation-as the features exemplified by a Z-representation aren't transferred to a target *unaltered*. In her discussion of imputation Elgin posits that a representation imputes the exemplified features "or related ones" to the target (2010, p. 10). This observation is particularly pertinent in scientific contexts. The features of a model are rarely, if ever, taken to hold directly in their target systems and so the features imputed to targets may diverge significantly from the features exemplified in the model. As an example consider the circulation period, the time it takes for the active balance to circulate once around economic system. As we have seen in the Sect. 8.2, there is a time-scale in the model with 1 min of machine running time being equivalent to 1 year in the economy. One can observe how long it takes for water to move around the machine once and then convert the machine time into economy time. Assume that it takes the drop 17 s to move around the machine. It then follows that it takes active balances 3.4 months to circulate around the economy. However, economists will not impute this precise figure to the particular economy that they take the machine to be a representation-of. When considering circulation time Phillips in fact reports that, according to the machine, the circulation period for the UK and the US economies seems to "somewhere between 3 and 4 months" (Phillips 1950, p. 291). So we have to transform "3.4 months" to "between 3 and 4 months" to get an imputable feature.

This is in line with Elgin's proposal because there is an obvious relation between "3.4 months" and "between 3 and 4 months". Nevertheless, the proposal needs some work. There is nothing incorrect about invocation of "related" features, but it lacks specificity. Any feature can be related to any other feature in some way or another, and as long as nothing is said about what this way is, it remains unclear what features are ascribed to T. One could put faith into context and hope that context will always determine what features are imputed to the target. We'd rather not. It remains unclear what a model says about its target as long as the relation between the features exemplified by the model and those imputed to the target is left

8.4 Imputation and Keys

unspecified. We therefore prefer to write such a specification explicitly into the definition of representation-as. Let $Z_1, \ldots, Z_m$ be the Z-features exemplified by the model, and let $Q_1, \ldots, Q_l$ be the "related" features that the model imputes to T (m and l are integers that can but need not be equal). The representation must then come with a key K specifying how the $Z_1, \ldots, Z_m$ are converted into $Q_1, \ldots, Q_l$.

Key K: let $M = \langle X, I \rangle$ be a model and let $Z_1, \ldots, Z_m$ be Z-features exemplified by M. A key K associates the set $\{Z_1, \ldots, Z_m\}$ with a set $\{Q_1, \ldots, Q_l\}$ of Z-features that are candidates for imputation to the target system. We write $K(\{Z_1, \ldots, Z_m\}) = \{Q_1, \ldots, Q_l\}$.

The third clause in the definition of representation-as then becomes: X exemplifies $Z_1, \ldots, Z_m$ and imputes some of the features $Q_1, \ldots, Q_l$ to T where the two sets of features are connected to each other by a key K.

The idea of a key comes from maps, which are paradigmatic for understanding scientific representation.[16] Consider a map of the world. It exemplifies a distance of 29 cm between the points labelled "Paris" and "New York". The map comes with a legend, which includes a scale, 1:20,000,000 say, and this allows us to translate a feature exemplified by the map into a feature of the world, namely that New York and Paris are 5800 km apart. The map's legend provides the key. Or consider the case of a scale model of a ship being used to represent the forces an actual ship faces when at sea. The exemplified feature in this instance is the resistance the model ship faces when dragged through a water tank. But this resistance doesn't translate into the resistance faced by the actual ship in a straightforward manner. The relation between the resistance of the model and the resistance of the real ship stand in a complicated non-linear relationship because smaller objects encounter disproportionate effects due to the viscosity of the fluid. The exact form of the key is often highly non-trivial and emerges as the result of a thoroughgoing study of the situation.[17] Determining how to move from features exemplified by models to features of their target systems can be a significant task, which should not go unrecognized in an account of scientific representation.

K is a blank to be filled. The key associated with a model depends on a myriad of factors: the scientific discipline, the context, the aims and purposes for which the model is used, the theoretical backdrop against which the model operates, etc. Building K into the definition of representation-as does not prejudge the nature of K, much less single out a particular key as the correct one. In some instances the key might be identity: the features exemplified by the model are imputed unchanged to the target. This could happen, for instance, if economists impute the fact that the model-economy destabilises in response to uncoordinated policy interventions unaltered to the UK economy. In other cases, the relation between the properties of the model and those imputed to the target may be similarity (see Chap. 3), which

---

[16] See Frigg's (2010b). The analogy between maps and scientific representation has been noted by a number of authors. For a discussion see Sismondo and Chrisman's (2001).

[17] See Pincock's (2019) and Sterrett's (2006, 2020) for a discussion of scaling relations.

can also ground an analogy (Hesse 1963). In yet others, the key might take the form of an "ideal limit" (Laymon 1990; Frigg 2010b, pp. 131–32), an option that we discuss further in Sect. 9.3.[18] But keys might also associate exemplified features with entirely different features to be imputed to the target (e.g., colours with tube lines as is the case in the London Underground map). At the general level, the requirement merely is that there must be *some* key for something to qualify as a representation-as; there is no requirement that it is of a particular kind.

8.5 The DEKI Account of Representation

We can now tie the loose ends together and give a full statement of the DEKI account. Those who adopt a universalist stance will see DEKI as a solution to the ER-Problem and will formulate the DEKI conditions as filling the blank in the ER-Scheme. We want to take a more cautious route here and proffer DEKI only as response to the Indirect Epistemic Representation Problem. This is because we find the distinction between direct and indirect epistemic representations (Sect. 3.1) appealing, and because DEKI seems to offer a natural answer to how indirect epistemic representations work, while its credentials as an analysis of direct descriptions (such as Darwin's account of atoll formation) seem to be rather questionable. Those who wish to uphold universalism can simply cross out the word "indirect" in front of "epistemic representation" and thereby recover a formulation of DEKI that is a response to the ER-Problem:

DEKI: let $M = \langle X, I \rangle$ be a model. M is an indirect epistemic representation of T iff M represents T as Z, whereby M represents T as Z iff all of the following conditions are satisfied:

  (i) M denotes T (and in some cases parts of M denote parts of T).
  (ii) M I-exemplifies Z-features $Z_1, \ldots, Z_m$.
  (iii) M comes with a key K associating the set $\{Z_1, \ldots, Z_m\}$ with a set of features $\{Q_1, \ldots, Q_l\}$: $K(\{Z_1, \ldots, Z_m\}) = \{Q_1, \ldots, Q_l\}$.
  (iv) M imputes at least one of the features $Q_1, \ldots, Q_l$ to T.

We call this the DEKI account of representation to highlight its defining features: denotation, exemplification, keying-up, and imputation. Notice that, in line with our discussion at the beginning of the chapter, here we have formulated DEKI as it applies to scientific models. However, it is clear that, and how, the account applies to indirect epistemic representations in general.

The interpretation is formulated in a certain language, namely the language we use when saying, or writing down, sentences like "one litre of water corresponds to a million of the model currency". And even before we write down an interpretation, the

---

[18] This allows a representation to be accurate even if it is *literally* false. For a discussion of literalism see our (2019b).

8.5 The DEKI Account of Representation

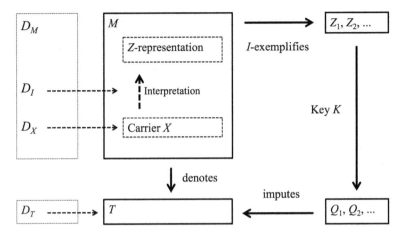

Fig. 8.4 Schematic representation of the DEKI account

concepts of the Z-domain have to be given to us through a language and the carrier has to be described in certain way. If we don't describe the Phillips–Newlyn machine as consisting of pipes and as having flows, then we cannot state the interpretation. We will often also describe the target in certain ways, not least when denoting it (more about denotation below). When dealing with material models the involvement of language is both obvious and unproblematic – or at least not more problematic than the use of language to describe objects in any other context. This will change, however, when we discuss non-material models in the next chapter, and to pave the ground for this discussion it is helpful to make descriptions a part of our statement of the DEKI account. To that end, let D_X be a description of X. In our example it would be the sort of description that Phillips gave of his machine *qua* machine. That is, D_X is couched entirely in a language consisting of terms referring to X-objects and X-features. Furthermore let D_I be a statement of the interpretation, which – obviously – consists of sentences that contain both X-terms and Z-terms. D_X and D_I taken together form the *model-description* D_M of the machine. Finally let D_T be a description of the target. There is no expectation that this description be rich or complete; it can be as minimal as pointing to an object and saying "this is my target".

Taking these elements together, we can represent the DEKI account schematically as shown in Fig. 8.4.

We can now present a complete analysis of how the model that has guided us through this chapter works. The Phillips–Newlyn machine, X, is used as the carrier of a model, and Z is an economy. The machine is endowed with Phillips' interpretation, I, mapping X-features to Z-features. The machine so interpreted is an open-IS-LM-economy-representation, and as such it is a model M. In the situation in which Phillips used the machine as a representation-of the UK economy he got the machine to denote the UK economy (i). He did so by borrowing the denotation of the linguistic expression "the UK economy" and the model ended up denoting whatever that expression denoted. The machine instantiates a number of water-pipe-features and, via I, it I-instantiates a number of economy-features. Some of them – the circulation period, for instance – are exemplified because they were highlighted (ii). Phillips then used an interval key, which moved from a specific value to an interval around it (iii) and he imputed the property of having a circulation period in that interval to the UK economy (iv).

This, we claim, is the right analysis not only of how the Phillips–Newlyn machine represents; it is also the right analysis for all other material models. The use of plasticine sausages as models of myoglobin, the use of mice as models for humans, etc. can all be analysed in terms of DEKI.

A few qualifications are in order. First, neither $\{X_1, ..., X_n\}$, nor $\{Z_1, ..., Z_n\}$, nor indeed $\{Q_1, ..., Q_l\}$, have to be sets of independent monadic properties. As made clear previously, by using the term "feature" we are highlighting that the sets can contain properties of any arity and at any level of abstraction, and features can depend on one another. In fact, the sets can be highly structured, for instance with some features expressing relationships between other features.

Second, a representation is accurate if T indeed possesses the features that M imputes to it. This need not be the case. As we have seen in the previous section, imputation is feature ascription. But T need not possess any of the features that are ascribed to it, and *no* assumption that it does is built into the notion of representation-as. M can represent T as possessing features $Q_1, ..., Q_l$ and T might not instantiate a single one of them. If M represents T as having features that it doesn't have, it misrepresents T with respect to those features. If T indeed has the features imputed to it by the representation, then the representation is faithful, again with respect to those features. Yet representations can be unfaithful.

Third, as previously noted, "scientific model" is not a synonym for "scientific representation". While all models are Z-representations, not all models represent *a target* as a Z, and, in fact, not all models are even a representation-of a target. Models featuring four-sex populations are four-sex-population-representations, but they are not representations-of anything. So our claim concerning scientific representation is not that all models are representations-of targets. Our claim is conditional: if a model is a representation-of a target, then it represents the target in the sense of DEKI.

Fourth, the ordering of the conditions in the definition of DEKI is a logical order and is not supposed to introduce a temporal element into the process of constructing and using a model. None of the conditions has to be established prior to the others. A model could, for instance, exemplify certain features even before being used as a representation-of a target by a scientist. Or the user could start with the target system and a set of features of interest. She could continue to construct an "inverse key", associating those features with ones that we have a firmer grasp on in the particular context of model building, and only then construct a model that exemplifies those features in the appropriate manner under the appropriate interpretation, before taking the model and establishing the denotation relation between it and the target. Such a process is not ruled out by our conditions. DEKI does not function as

8.5 The DEKI Account of Representation

a diachronic account of scientific representation: as long as the conditions are met, in whatever order, a model represents its target system as Z.

Fifth, DEKI is the general form of an account of epistemic representation and as such it needs to be concretised in every particular instance. In every case of a carrier representing a target one has to specify what X is, how it is interpreted, what sort of Z-representation it is and what features it exemplifies, how denotation is established, what key is used, and how the imputation is taking place. Depending on what kind of representation we are dealing with, these "blanks" will be filled differently. But far from being a defect, this degree of abstractness is an advantage. "Scientific modelling", and indeed "epistemic representation", are umbrella terms covering a vast array of different activities in different fields. A view that sees representations in fields as diverse as macroeconomics, biochemistry, and fluid dynamics, and indeed beyond science in cartography, installation art, and cave painting, as being covered by the same account must have some flexibility in the account to make room for differences between representations. Our definition occupies the right middle ground between generality and specificity: it is general enough to cover a large array of cases and yet it allows us to say what is specific about them. The fact that DEKI is the general form of an account of (indirect) epistemic representation becomes particularly palpable in two places. The first is the key. The definition of a key we stated is abstract and in every particular case of epistemic representation a specific key will have to be chosen. We have seen how certain features of the Phillips–Newlyn machine are endowed with a key, and we will discuss in some detail in the next chapter how limits can be used to define keys. The definition of DEKI neither provides nor constrains keys. The second place is the interpretation, where scientists enjoy great freedom of choice.

Sixth, as with images, that a model as a whole denotes a target as a whole does not preclude there being additional denotation relationships between parts of the model and parts of the target. The Phillips–Newlyn machine as a whole denotes the UK economy, and parts of it – for instance the tank labelled "foreign owned balances" and the flow labelled "income" – denote parts of the economy. This is accounted for in the parenthetical qualification in condition (i). This raises the vexing question of how denotation is established. In many cases denotation is borrowed from language. In a map of London we see a black dot with "Holborn Station" written next to it. The dot borrows denotation from language and thereby comes to denote Holborn Station. Many models seem to work in the same way. The Phillips–Newlyn machine denotes whatever the words "UK economy" or "Guatemalan economy" denote. At least in cases where this happens, this is to hand over the problem of uncovering the roots of denotation to the philosophy of language. In the philosophy of language there are two broad families of approaches.[19] According to the descriptivist approach (which goes back to Frege and Russell), names function as disguised definite descriptions, and as such denote whatever

---

[19] For discussions and surveys see Lycan's (2000, Chaps. 4 and 5) and Michaelson and Reimer's (2019).

satisfies them. According to the so-called direct reference approach (which goes back to Mill, Barcan Marcus and Kripke), names directly pick out their bearers without going via any descriptive content. Both of these are in principle compatible with our view of scientific models and for now we want to remain agnostic about this choice. We also have pluralist leanings and want to make room for there being different ways to establish denotation in different cases and in different contexts.

Let us now turn to the problems and conditions that we formulated in Chap. 1 and discuss how DEKI deals with them. DEKI gives a negative answer to the Scientific Representational Demarcation Problem. For those who reject the distinction between direct and indirect epistemic representation it also offers a negative answer to the Taxonomic Representational Demarcation Problem; for those who accept the distinction it demarcates taxonomically between direct and indirect epistemic representations and puts forward the DEKI conditions as an account of the latter. As noted previously, the DEKI scheme itself does not offer specific keys, nor does it determine how denotation is established. DEKI does therefore not offer a concrete response to either the Problem of Style or the Problem of Accuracy. Yet, and that's a crucial point, it creates a systematic space in which these considerations can be discussed. For example, with respect to the Problem of Style, it allows for style to be a multi-faceted aspect of scientific practice: some modelling styles depend on the kind of objects used as the carrier (one might identify the model-organism-style); other styles might depend on the sorts of interpretations used (economy-representations might thus form a style); and yet others might depend on the kinds of key used (the limit-style models discussed in the next chapter can be seen as forming a style). DEKI does address the Problem of Carriers explicitly. We have seen in this chapter how DEKI deals with material models, and we will see in the next chapter how it deals with non-material models.

In response to the Surrogative Reasoning Condition, DEKI offers a clear account of how we generate hypotheses about the target from the model: we figure out which features are exemplified, convert these into other features using the key, and then impute these other features to the target. The Misrepresentation Condition is met because, as noted previously, there is no expectation that T actually instantiate the features that M imputes to it. Furthermore, misrepresentation can also enter via the denotation condition. Denotation can fail in various ways: a representation can purportedly denote a target that does not exist, or it can denote the wrong target. DEKI only requires that the model denotes; it does not require that it denotes successfully or correctly. The Targetless Representations Condition is met by construction, because Z-representations don't have to be representations-of anything and yet they are representations. The Directionality Condition is met because models denote targets and not vice versa, and because models generate hypotheses about targets and not vice versa. The Applicability of Mathematics Condition has not been addressed so far. We will do this in the next chapter when we discuss DEKI in the context of non-material models.

8.6 Reverberations

In this section we discuss a number of issues that arise in connection with DEKI and indicate where we stand on these.

One might worry that DEKI is a similarity account in disguise, where the disguise consists in replacing similarity with something we call a "key". But, so the worry goes, a key is similarity in all but name, and so the two accounts are really just notational variants of each other. This is not so. While a key *can* be based on a similarity relation (as Elgin already noted in the passages quoted at the end of Sect. 7.6), there is no assumption that this always has to be the case and hence introducing keys does not amount to smuggling a mimetic conception of representation into the account. Keys can be highly conventional. After a short immersion in a solution, a strip of litmus paper exemplifies a certain shade of red, and, via a key that converts a colour spectrum into levels of acidity, ascribes a pH value of 3.5 to the solution. But redness is not similar to acidity.

As another example of conventional key consider the representation of the Mandelbrot set in Fig. 8.5 (readers can find a colour version of the figure on the website indicated in the caption to the figure). The image is generated by using a key

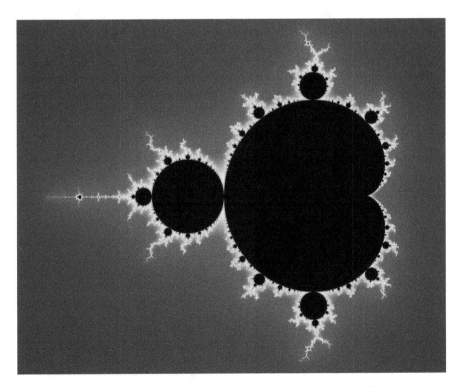

Fig. 8.5 Mandelbrot set. The figure is from https://commons.wikimedia.org/wiki/File:Mandelbrot_set_10000px.png and reproduced here under the CC BY-SA licence

that translates divergence speed into colour. The rectangle shown is a segment of the complex plane and each point represents a complex number c. This number is used as parameter value for an iterative function. If the function converges for c, then the point in the plane representing c is coloured black. If the function diverges, then a shading from yellow to blue is used to indicate the speed of divergence, where yellow is slow and blue is fast (Argyris et al. 1994, p. 663). This representation does not fit the mould of the similarity account because there is no similarity relation between colours and divergence speeds.

Keys being conventional is not the same as keys being voluntary. Scientists enjoy complete freedom in choosing their keys, but they don't enjoy the freedom of dispensing with the key altogether: no key, no epistemic representation! If a model is so far removed from anything in the world that scientists don't know how to specify a key, then they don't have an epistemic representation of anything in the world. Such a model can still be a Z-representation, but it is then not a representation-of a target. The construction of a key to accompany a scientific model is an important part of scientific practice: if one's preferred model is be used to provide information about an actual system in the world, then one has to be explicit about how features of the model are supposed to correspond to features one conjectures the target to have. Ignoring this connection amounts to investigating the model without investigating any actual system in the world.[20]

A consequence of our account is that there are two different notions of content in a representation, which we refer to as manifest content and representational content respectively. The *manifest content* is the content of the Z-representation. The manifest content of a dragon-picture is a dragon with all the details that the picture contains; the manifest content of an economy-representation is an economy; and so on. In our account the manifest content of a Z-representation depends on the carrier and the interpretation, and the complete manifest content of the Z-representation is effectively given by the features that are exemplified by X under an interpretation I. This content is distinct from the *representational content*, which is the totality of what the representation says about the target. In our account, the representational content is everything that the representation imputes to the target, which results from the exemplified features and the key. In other words, the representational content is the subset of $\{Q_1, ..., Q_l\}$ which consists of the features that are actually imputed to T. It is obvious that these two notions of content are distinct because a Z-representation need not impute anything on a target system, in fact it need not have a target at all.

On closer examination it turns out that the notion of representation-as that we have developed in Chaps. 7 and 8 is to some extent misaligned with the pre-theoretical use of the phrase "representation as". One is tempted to say things like "the

---

[20] In practice there is often a division of labour involved. Theoreticians may prefer to focus on investigating their models independently of providing them with a key. This is valuable in and of itself, since it furthers our knowledge of the models we use. But if a model is to be used to tell us something about the world – in a policy setting for example – then practitioners will have to use a key.

8.6 Reverberations

Phillips-Newlyn machine represents the UK economy as a system of reservoirs and pipes". According to both Representation-As and DEKI this would be wrong. The source of the mismatch is a conflation of Z and X. On the account developed in this chapter, a model $M = \langle X, I \rangle$ represents a target T as Z, where the Z is introduced through the interpretation; it does not represents T as X. However, the system of reservoirs and pipes in the Phillips–Newlyn machine is the X of the model and not the Z. In the Phillips–Newlyn machine Z is an open-IS-LM economy and under the standard interpretation the machine is an open-IS-LM-economy-representation. On our account, then, the machine represents the UK economy as an open-IS-LM economy, and not as a system of reservoirs and pipes. A correct paraphrase of the pre-theoretical claim would be something like "the Phillips-Newlyn machine proffers a system of reservoirs and pipes as the carrier of the representation of the UK economy". We recognise that this stands in tension with a pre-theoretical use of the expression "representation as", but philosophical – or indeed scientific – explication often comes at the costs of misalignment with everyday notions: recall the discomfort that zoology causes common sense when it declares whales not to be fish.

Whether a representation is good or bad, or useful or useless, depends on the context of the investigation and on the aims, questions, and objectives that scientists pursue. There is no such thing as an "intrinsically" or "absolutely" good or bad representation. An account of representation (DEKI or otherwise) has to explain how something represents something else, but it does not at the same time have to state criteria of goodness. Whether a given representation serves certain purposes or not is a question that stands outside an account of representation. A map can be good for one thing but not for another. A road atlas is useful when driving from London to Glasgow; if we then go hiking in the Scottish Highlands with the same road atlas we are likely to lose our way. The liquid drop model of the nucleus is useful to calculate fission energies; it leads us astray when we want to understand the inner structure of a nucleus. Representations represent what they do, and an account of representation has to tell us how they do so. Such an account doesn't have to also issue warnings about what we can and cannot do with certain representations, or warn us about when these representations are misleading. When and to what extent a representation can be trusted is an important question, but it is not one that we should expect to be answered by an account of representation.

It also ought to be noted that questions concerning representation are orthogonal to the scientific realism issue. Even when models have all kind of unobservable properties, models can be interpreted as exemplifying only observable properties, or only observable properties can be keyed up and imputed. Such an understanding of a model is antirealist. If all features, also unobservable ones, are exemplified, keyed up, and imputed, then a model is understood realistically. Sometimes instrumentalism is presented as the view that models don't represent. This is wrong. From an instrumentalist point of view models should not be thought of as representing unobservable features or mechanisms, but even an instrumentalist requires that models represent observables. A model can only be an "inference ticket" when it represents observable features and connects these to other observable features in the model via some mechanism. The mechanism itself need not be keyed up and hence need not represent; but the observable features that are connected through the mechanism do have to be taken to represent – if they did not, the model would lack a connection to reality and not be instrumentally useful.

Finally, DEKI offers an interesting angle on perspectivalism and the "problem of multiple models". As discussed by Giere (2006) and Massimi (2017, 2018) (see also the references therein, and the references in footnote 3 of the Introduction), in various areas of modern science, scientists use multiple models of the same target system. How are we to understand this pluralistic aspect of scientific practice? From the point of view of DEKI the pertinent points are the following. When a model represents a target T as Z it imputes features $Q_1, \ldots, Q_l$ to T. There is no supposition that these features exhaust the features of T; in this sense models provide only partial perspectives on their targets. However, in the literature on perspectivalism there is an additional worry: it's not just that each model represents a different feature of the target, rather the models appear, at least prima facie, like they provide mutually *inconsistent* representations of the same target. The question that arises then is what it means for two models to be inconsistent with one another. From the point of view of DEKI the natural notion of inconsistency that arises concerns cases where one model imputes a feature Q_i and another feature imputes Q_j, where these features are such that nothing can have both at the same time. So the question concerns the consistency of the models' representational content. But notice that this isn't the same as the models themselves having inconsistent manifest content: neither a shell model nor a liquid drop model of the nucleus has to impute being a shell or a liquid drop to their shared target system: the fact that the models exemplify their respective features, which might be inconsistent with one another, does not entail that the models represent their targets inconsistently. Indeed, the fact that some scientists use different models successfully may very well have an impact on how other scientists key up their preferred models. The addition of a key allows for some flexibility, and it's an interesting avenue for future research to inquire into how the DEKI machinery can address, or dissolve, the questions that arise in connection with multiple models of the same target system.

Chapter 9
DEKI Goes Forth

In the previous chapter we developed the DEKI account of epistemic representation against the backdrop of a material carrier. We begin this chapter by exploring how it applies to carriers that are non-material (Sect. 9.1) and how it answers the Applicability of Mathematics Condition (Sect. 9.2). We then further investigate the nature of keys by discussing an important class of keys based on limits (Sect. 9.3). Finally, we examine DEKI's stance on the demarcation problems with a comparison of epistemic representation in art and science (Sects. 9.4 and 9.5).

9.1 DEKI with Non-material Models

The DEKI account explains how an object becomes first a Z-representation and then represents a target system as thus and so.[1] As developed in the previous chapter, the account relied on the materiality of the carrier: as a material object it is able to instantiate material properties. But many scientific models are not material objects, which, as we have seen in Sect. 3.4, renders the notion of property instantiation tenuous. This raises the question how the DEKI account can be formulated so that its conditions are met even when carriers don't instantiate properties in the way in which material objects do; that is if they don't themselves possess physical features.

It is worth noting at the outset that the DEKI account enjoys a considerable advantage over similarity accounts because it does not require carrier features to function representationally tout court. Rather, the account requires features that are *I*-instantiated, i.e. instantiated under an interpretation, and this makes a significant difference. The Phillips–Newlyn machine does not have to literally instantiate economic features to be an economy-representation, or to represent an economy,

---

[1] In this section we focus on non-material models rather than non-material carriers in general. See Sect. 9.4 for further discussion of non-material carriers in non-scientific contexts.

which is something it would never be able to do. All that is required is that the machine instantiate some features that can be *interpreted* in terms of economic features. The account retains this flexibility when we shift to non-material models. The account does not require a model to instantiate the features that it imputes (or, indeed, the features that are fed into the key). It does not require, for instance, the Newtonian model to be such that model-planets really have mass and really move in trajectories, which is something that model-objects will never do. All that it requires is the model has some feature or other that can be so interpreted. This makes the DEKI account immensely flexible when it comes to ontology: in principle anything that has any feature that we can interpret in terms of another feature can be a model. In particular, there is nothing in DEKI per se that would rule out set-theoretic structures, and DEKI could in principle be used to articulate a theory of representation that follows the structuralist account in taking models to be set-theoretic structures.

We will return to how a structuralist could use DEKI at the end of this section. For now we want to take a different route and see how DEKI connects to the fiction view of models developed in Chap. 6. As we have seen in Sect. 6.2, approaches belonging to the first branch develop the fiction view as a response to the Problem of Carriers and leave the issue of representation unresolved. This omission can now be rectified by combining the first-branch-fiction-view with the DEKI account of epistemic representation (from now on we only speak of the "fiction view" and take it to be understood that we are referring to the first branch of the view; since the second-branch-fiction-view is already an account of representation it cannot be combined with another account of representation, DEKI or otherwise).

Combining DEKI with the fiction view makes use of the above-mentioned flexibility of DEKI as regards carriers. Since nothing in the account depends on carriers being any specific sort of object, let alone a physical object, we are free to admit fictional objects as carriers. This is the leading idea: simply put a fictional object in the place of the material object in the schema developed in Sect. 8.5! The details of how this is done depend on what account of fiction one adopts. As we have seen in Sect. 6.6 the currently most popular versions of the fiction view, both realist and antirealist, are developed using pretence theory. So we formulate our combination of DEKI with the fiction view using a version of the fiction view based on pretence theory – without, at this point, committing to either a realist or an antirealist interpretation of the products of pretence.[2]

The key to fusing the fiction view with DEKI is to look at the role of descriptions. Looking at Fig. 8.4 in Sect. 8.5 we see that the DEKI account features a model description that consists of two parts, a description D_X of the carrier and a description D_I of the interpretation that turns the carrier into a Z-representation. Now compare this to pretence theory. In Sect. 6.5 we have seen that non-material models are given to us through model-descriptions, which serve as props in a game of make-believe. Furthermore, like the model-descriptions of a material model, these

---

[2] An alternative way of unifying material and non-material models is to also bring material models under the umbrella of pretence theory. For a discussion of this alternative see Toon's (2011) and our (2016b, p. 236).

9.1 DEKI with Non-material Models

model-descriptions also have components that are analogous to D_X and D_I. Take again the example of Newton's model of a celestial orbit. D_X generates the carrier, namely what is usually called the *two-body system*: a system consisting of two homogeneous perfect spheres, one large and one small, attracted to each other with a $1/r^2$ force. D_I instructs us to imagine the larger sphere as the sun, the smaller sphere as the earth, and the force as gravity. Keeping D_X and D_I separate is not just a contrivance of philosophical analysis. In fact, history testifies to their distinctiveness. Consider the Bohr model. That model leaves D_X intact but replaces Newton's solar system D_I with a hydrogen atom $D_{I'}$ which instructs us to imagine the large ball as a proton, the small ball as an electron, and the force as Coulomb attraction.

So we get from the DEKI account for material models to the DEKI account for fictional models by reconceptualising the D_X part of model-descriptions. In the context of a material model D_X is a description of the material object; in the context of a fictional model D_X is a prop in a game of make-believe that mandates the scientist to imagine the content of the description. D_I remains the same: it relates X-features to Z-features, the only difference being that in a fictional model X-features pertain to fictional objects while in material models they belong to material objects.

There is, however, one crucial difference between material and fictional models. Models are objects that behave in a certain way and have an internal dynamic that scientists study. In the case of a material model this dynamic is the dynamic of the carrier object. The Phillips–Newlyn machine pumps water in a certain way, reacts in a specific manner to closing a valve, does particular things in response to a readjustment of certain levers, and so on. Scientists find out what the machine does by performing an experiment on the machine. In the case of a fictional model, one can't just switch on the model and see what it does. The internal dynamic of the model is not given by its material mechanism, but by the principles of generation. D_X will operate against the background of such principles which allow those involved in the game to reach conclusions that have not been written into the basic specification of the model. For instance, that the small sphere moves in an elliptical orbit around the large sphere is a proposition that is true in the two-body game of make-believe but does not form part of the basic specification of the two-body system. In this way an imagined-object can play the same role in a non-material model as a concrete object in a material model. We explore the behaviour of a fictional model by finding out what follows from the basic model assumptions stated in D_X when combined with a set of principles of generation. To do so we use the principles in a certain formulation. Let D_G be a description of these principles, which, for the reasons we have just seen, should be considered to be part of the model-description. So the crucial difference between material and non-material models is the inclusion of D_G in the model description. Integrating this change into the DEKI picture leads to Fig. 9.1.

Comparing Figs. 8.4 and 9.1 makes clear how similar the two versions of the DEKI account are. The only thing that has changed is the inner constitution of the model-description (we added D_G) and the understanding of D_X (which is now seen as a prop in game of make-believe rather than as a description of a physical object), and the fact that the carrier X is now a fictional object rather than a physical object.

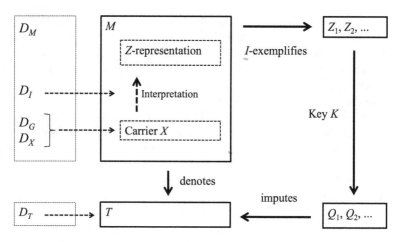

Fig. 9.1 The DEKI account for fictional models

We take this uniformity to be an advantage, and that DEKI can cover both kinds of models with only minor modifications is one of its selling points.

The advantages of pretence theory now become palpable. The account needs a carrier with features "of its own" to explore to be useful, and this is what pretence theory provides. As we have seen in Sect. 6.5, scientists often start with a few basic posits, make certain assumptions, and rely (often tacitly) on background theories. By actively manipulating these elements under certain constraints, and by seeing what fits together and how, they learn about the posits and about what they imply. This activity is naturally analysed as being involved in a game of make-believe, and in doing so an imagined-object is explored. Due to the use of principles of generation the imagined-object can have properties and features that have not been written into the original model-description, which is why the study of imagined-objects is cognitively relevant. By being involved in such games, physicists learn about the geometrical properties of orbits and population biologists about growth of populations, neither of which were explicitly mentioned in the model-descriptions. Nothing in DEKI depends on there being a real object that literally instantiates a physical feature. What matters is that there is right and wrong in feature attribution, and make-believe explains constraints to imagination cogently in terms of facts about the prop and adherence to principles of generation. So pretence theory offers everything the account needs.[3]

As in the case of material models the interpretation need not, but can be, couched in X's own terms. A ship-like carrier can be interpreted in ship-terms, which makes it a ship-representation. In the case of fictional models this means that D_X and D_I

---

[3] In passing we note that the comparison problem we encountered in Sect. 6.5 does not arise in this version of DEKI because the account does not require comparative claims. The fourth condition in DEKI is that features are imputed to the target. In linguistic terms this means that claims of the form "target T has feature Q" are put forward. These are standard attributive claims rather than comparisons, and as such they raise no problems having to do with fiction.

9.1 DEKI with Non-material Models

will be couched in the same terms. In Fibonacci's famous model, for instance, the imagined-object specified by D_X is a rabbit population and the model is a rabbit-population-representation. So the interpretation part is reduced to identity (but not so the key: the features of Fibonacci's fictional rabbits are not imputed unchanged to real rabbits). Interpretations of this kind seem to be much more common in fictional models than in material models. This is because the imagination is less constrained than the material world and so it's often easier to find a suitable fictional object than to come by an appropriate material system.

The DEKI account as applied to fictional objects inherits most of the responses to the problems and conditions in Chap. 1 from the version with material models in Sect. 8.5, and we will not repeat them here. There are, however, two conditions that are handled differently: the Directionality Condition and the Applicability of Mathematics Condition (which has not been addressed at all for material models). Before turning to mathematics in the next section, it is worth briefly commenting on the Directionality Condition. DEKI meets the condition through its requirement of denotation: models denote targets and not vice versa. However, as we have seen in Sect. 6.6, denotation is a dyadic relation between certain symbols and certain objects and if carriers don't exist they can't denote. In the context of material models this was not a problem. But in the context where the carriers in question are fictions things look slightly different. Fictional realists can grant that carriers exist, and therefore are the sorts of things that can denote. Fictional antirealists, however, have to provide an account of how fictional carriers can denote. Some fictional antirealists might argue that pretend denotation is enough to introduce the kind of asymmetry required to meet the Directionality Condition. Those, like Salis (2019), who think that representation requires real denotation need to do more work. This is what prompted Salis to formulate her new fiction view of models. The main departure of the new view from the old view is that the new view associates the model not with the imaginings of scientists but with the model-description and its content. This change can be implemented in the above by drawing the box called M around the model-descriptions rather than the carrier, and by somehow making explicit that the content of the description is also part of the model. The details of this require some work, but they present no fundamental obstacle.[4]

So far we have outlined how to fuse DEKI with our preferred answer to the Problem of Carriers: non-material carriers are fictional objects that play a representational role via the DEKI conditions. However, as mentioned at the beginning of the section, the DEKI account itself is flexible regarding the nature of carriers. Another option is to fuse the DEKI account with the idea that carriers are mathematical structures. Here the idea is the following. Take non-material model carriers to be mathematical structures of the sort discussed in detail in Chap. 4, and then combine them with an interpretation. This turns a mathematical structure into a Z-representation; and if the structure has been chosen by a scientist (or group of

---

[4] The necessary moves are spelled out explicitly in Salis et al. (2020).

scientists) to be the carrier of a model, then that Z-representation is a model.[5] This approach can be accommodated exactly as shown in Fig. 9.1. Here, the X, rather than being a material object is a mathematical structure. D_X and D_G specify the mathematical structure in question (and thus some of the issues discussed in Sect. 4.2 do reappear here). X has mathematical features, which are interpreted as physical features by D_I. Thus the model system can be thought of as an interpreted mathematical structure, which then represents some target system T via the remaining DEKI conditions.

These interpreted structures are analogous to what Weisberg (2013, Chaps. 3 and 4) also calls "interpreted structures". According to Weisberg, we should think about models as composed of two parts: a "structure" and a "construal".[6] The latter includes four parts: an "assignment", an "intended scope", and two "fidelity criteria".[7] Of these, the most important is the assignment, which consists of "explicit specifications of how parts of real or imagined target systems are to be mapped onto parts of the model" (ibid., p. 39). The difference between what we're suggesting and Weisberg's discussion is that our D_I-interpretations can be given without reference to any particular target system. They provide a physical interpretation of mathematical features, but these interpretations concern physical features in general, without making essential reference to how these features may or may not appear in systems out there in the world. Thus one can talk about the population in the Fibonacci model reproducing at a certain rate because a mathematical feature is interpreted as the physical feature *reproduction* without thereby representing any actual population as reproducing at that rate (indeed, as mentioned previously, if one were to use the model to represent a particular target system, one wouldn't use an identity key). The advantage, then, of thinking about interpreted structures in terms of Z-representations is that we can make sense of the practice of talking about mathematical models in physical terms without requiring that the physical features described are realised, or thought to be realised, in a particular way in any actual target system.

However, despite this difference, in the context of fusing together a mathematical answer to the Problem of Carriers with the DEKI account of epistemic representation, it's important to note that all the benefits of Weisberg's "assignments" are also captured by our notion of an interpretation. As applied to a mathematical object, an

---

[5] Notice that the way in which a Z-representation then represents a target as thus and so does not need to proceed via the structuralist view of epistemic representation. In fact, there is then a clear division of labour: DEKI handles the epistemic representation and structuralism handles the Problem of Carriers, and, as a result, many of the issues discussed in Chap. 4 don't arise.

[6] Weisberg distinguishes between concrete, mathematical, and computational structures. It's the mathematical and computational structures that are relevant here, and the distinction between the two is not relevant for our current purposes. For a discussion that applies the DEKI framework in the context of computational models, in particular the use of agent-based simulations in the social sciences, see Gräbner's (2018).

[7] The intended scope specifies which aspects of potential target phenomena are to be represented by the model, and the fidelity criteria specify the standards of accuracy for the model to be an accurate representation of a target system.

interpretation D_I, needn't interpret *every* feature of the mathematical structure as a physical feature. To borrow an example discussed by Bueno (2005) and Bueno and Colyvan (2011), when Dirac found negative energy solutions to (what we now call) the Dirac equation, they weren't initially regarded as physically meaningful aspects of the model. But after the discovery of positrons they were. This is easily accommodated by noting that it corresponds to expanding the interpretation D_I associated with the structure to include interpreting the mathematical feature associated with negative solutions in terms of positrons, particles that have the same mass as the electron but opposite charge. Or alternatively, returning to the discussion of "surplus structure" in Sect. 4.4, in electromagnetism we can provide interpretations of mathematical gauge fields that only associate physical features with the exterior derivative of a one-form, rather than features that vary across fields related by a gauge transformation. So not all aspects of a mathematical model need to be representationally relevant, in the sense of being provided a physical interpretation, and we can make sense of this aspect of mathematical modelling without reference to a target system.

9.2 The Applicability of Mathematics

Whilst the DEKI account combined with a structuralist answer to the Problem of Carriers might go some way to addressing the Applicability of Mathematics Condition, it is less obvious how the combination of a fiction view of models and the DEKI account does so. In this section we discuss how this combination of views addresses the condition. There are in fact two subtly distinct ways in which mathematics enters the picture according to the DEKI account. The first concerns the mathematisation of a target system; the second concerns the mathematisation of the carrier. We discuss these questions in order in this section.

Recall the discussion from Sect. 4.5 of where to find the target end structures required by the structuralist view of epistemic representation. Four options were sketched: target end structures are *data models*; targets are *ascribed* structures; targets *instantiate* structural universals; and, more radically, targets *are* structures in the sense that the world is considered to be a mathematical object. In principle, each of these options is available here too. The search for a target end structure was *required* by the structuralist view of epistemic representation, but if a successful answer can be given it needn't be tied to the structuralist view of representation. In fact, if a successful account of target end structures can be given, then the world can be represented in mathematical terms also based on the DEKI account: a mathematical feature Z_1 can be taken to be exemplified by the model, and another mathematical feature Q_1 (related to Z_1 by a mathematical key) is imputed to the target. The presence of a key liberates the DEKI account from thinking of the applicability of mathematics in terms of a structure preserving map between the mathematical structure of a carrier and the mathematical structure of a target (which ties in with the points in Sects. 4.3 and 4.4, which suggest that it explicitly shouldn't be thought of in these terms, at least if we are to account for misrepresentation and idealisation).

However, as we have seen in Sect. 4.5, there are serious issues with some of the accounts of target end structures, and in our (2017) we advocate the idea that target systems can be ascribed structures. The account applies here without alterations. Recall the idea that a target system is given a physical description, $\boldsymbol{D_S}$, that has the form: target T consists of objects $\boldsymbol{o_1}, \ldots, \boldsymbol{o_n}$ that either individually, or in collections, instantiate physical properties and relations $\boldsymbol{r_1}, \ldots, \boldsymbol{r_m}$.[8] $\boldsymbol{D_S}$ is a "structure generating description" in the sense that it generates a further description D_S, no longer in bold font, which is the extensional counterpart of $\boldsymbol{D_S}$, and describes the system as consisting of such a number of objects entering into such and such extensional relations. This describes a physical system in structural terms, thus providing a target end structure that can be represented according to one's preferred account of epistemic representation.

In the case of DEKI this works as follows. A key connects features $Z_1, \ldots, Z_n$ with features $Q_1, \ldots, Q_m$. Some of these features can be mathematical, for example the population growing *exponentially* up to a certain time when it begins to approach the environment's carrying capacity, or the planet moving at such and such a *velocity* (given by a real number) and in a thus and so shaped *orbit* (specified mathematically, say). This can be connected to DEKI in two ways. The first way locates the structure generating description in the imputed features themselves. We have imposed no restriction on the Qs and they can be as complex as we like. So Q_i, say, can be a complex property that is specified by $\boldsymbol{D_S}$, meaning that to impute Q_i to the target is tantamount to saying that description $\boldsymbol{D_S}$ is true of the target. The key can then specify that both Q_i and its "extensionalisation" are imputed, and in imputing the "extensionalised" version of Q_i the model ipso facto imputes the structure specified by D_S to the target.[9] Hence, the mechanism of ascribing a structure to a target system through a structure generating description can be connected to DEKI in a natural manner. The imputation is true if $\boldsymbol{D_S}$ is true of the target, which implies that also D_S is true of the target. If this is the case, the target can be said to have the mathematical feature described by D_S. It's important to note that, on this account, what makes the ascription of a mathematical feature to a target true is, strictly speaking, the fact that it has a physical feature which has that mathematical feature as its extensionalised counterpart.

The second way to connect the structure generating descriptions account to DEKI is to locate $\boldsymbol{D_S}$ in the description D_T of the target system (see Fig. 9.1; notice that D_T isn't being used here to signify an extentionalised description). A target system then has a mathematical feature F if the following holds: D_T incorporates the physical description $\boldsymbol{D_S}$; $\boldsymbol{D_S}$ is true of the target (which means that, a fortiori, D_S is true of the target); and the structure described by D_S has the feature F. Some of the

---

[8] Recall that the use bold font indicates that the terms in $\boldsymbol{D_S}$ refer to physical objects and relations. The terms individuating the $\boldsymbol{o}$'s are terms like "rabbit" not "element of a set", and the $\boldsymbol{r}$'s are terms like "reproduces with" not "being a symmetrical binary relation".

[9] We use the term "extensionalisation" and its cognates to refer to the result of the move from $\boldsymbol{D_S}$ to D_S.

9.2 The Applicability of Mathematics

features that the model imputes to the target can be mathematical, and if the target has these mathematical features in the sense just stated, then the imputation is true. Thus, a carrier can represent a target system as having a mathematical feature, the claim that the target has this feature can be true of the target system in virtue of the target system being accurately described by a physical description, which in turn can be converted into a description of a structure that has that mathematical feature.

It's crucial to understand how this is supposed to answer the Applicability of Mathematics Condition. In our (2017) we distinguish between the "general application problem" and the "special application problem". The former is the question: in virtue of what does *any* mathematical structure apply to a target system? The second is in virtue of what does a specific (type of) mathematical structure apply to a given (type of) target system? In this vein one can ask, for instance, in virtue of what the natural numbers apply to a rabbit population; in virtue of what manifolds apply to spacetime; in virtue of what Hilbert spaces apply to quantum systems; and so on. By appealing to structure generating descriptions we attempt to answer only the general application problem. Thus, we haven't addressed here the question of how a target system can have a specific mathematical feature (like having a particular differential structure for example). To do this would require a much more detailed story about specific applications of mathematics, requiring us to ask questions like "what is the physical basis for ascribing one particular mathematical structure to a target but not another?", which can be meaningfully discussed only on a case-by-case basis. Our discussion has clarified what it means, conceptually speaking, to apply a structural, and hence mathematical, feature to a physical target system, and this paves the ground for a discussion of the special application problem in specific cases.

So far we've discussed how a target system can be described in mathematical terms, and how this provides a target end mathematical structure that can then be represented by a carrier. The next question then is how we should think about carriers in mathematical terms. Let us begin with material models. As least in some cases, material carriers also have mathematical features that play a representationally relevant role. For example, we might say that the model-ship has a certain velocity, or faces a certain resistance, when travelling through the liquid in the model, and these are measured by real numbers. When we do this, we might also exploit mathematics to reason about those numbers (e.g., inferring the value of a quantity from the value of another quantity). How to think about the application of mathematics in such cases should now be clear. Material carriers, themselves being physical systems, can be subjected to exactly the same treatment as target systems: the material carrier can be described in physical terms (through D_X, see Fig. 8.4) this description generates a structure according to which the carrier has certain structural features (under a physical description). This then, allows the carrier to exemplify those features, and the rest of the representational machinery associated with DEKI can kick in.

The case of non-material carriers is more complex because how non-material carriers can have mathematical properties will depend on what one takes those carriers to be. As discussed in the previous section, DEKI is compatible with both the idea that carriers are fictional objects, and the idea that they are mathematical structures. If the carriers are taken to be mathematical structures then it's easy to see how mathematics enters the picture because carriers are mathematical objects.

The issue is a bit more involved if one takes carriers to be fictional objects. Recall that in the case of fictional carriers we have a set of descriptions D_X and D_G that prescribe us to imagine a fictional object with such and such features. They prescribe us to imagine, for instance, two perfect spheres that interact gravitationally. What does it mean to apply mathematics to such a carrier? We here discuss how mathematics fits into the account of carriers in terms of pretence that we discussed in Sects. 6.4 and 6.5, and we regard it as an open question how the mathematisation of fictional entities in alternative accounts of fiction would proceed.

Recall that the content of the fiction is generated in a game of make-believe based on D_X and D_G, the description of the carrier and the principles of generation. The clue to understanding mathematisation in this context is the realisation that nothing in our account of the applicability of mathematics in Sect. 4.5 depends on structure generating descriptions to be descriptions of a concrete object. In fact, any meaningful descriptive sentence couched in a concrete language can be a structure generating description, irrespective of whether it is put forward as description of a concrete entity or whether it is uttered in pretence. The description can be extensionalised in the same way in both cases, which leads to a mathematical structure. If the relevant sentence (or set of sentences) is (are) uttered in pretence, then the extensionalisation takes place in pretence and it is then part of the game of make-believe that we pretend that the object of our pretence has a certain mathematical structure. In other words, we can treat D_G and D_X as being structure-generating descriptions.[10] So when one is prescribed to imagine something physical in a game of make believe, one can also imagine its extensional counterpart.[11] In this way, a fictional object, even if described non-mathematically, can have mathematical features. One imagines some physical feature and, as per the structure generating procedure described above, one also imagines some mathematical feature.

However, a brief look at scientific practice shows that in many cases model descriptions will themselves be given in mathematical terms. Mathematical concepts can be part of descriptions or rules like the topography of a city can be part of a novel. So specifying a carrier can already involve mathematical concepts, for instance when we specify that a perfect sphere is part of the carrier. Thus, the language in which D_X is formulated can, and often does, contain mathematical terms. Likewise, the principles of generation can also contain mathematical rules and so D_G can, and often does, also contain mathematical terms. In Fibonacci's case basic arithmetic concepts are used in D_X and the principles of generation applied in the model contain full-fledged arithmetic, which is used to generate the population size numbers at later times (which are not part of the carrier description). In other cases, the principles of generation contain mathematically formulated laws of nature that

---

[10] Strictly speaking, then, the appropriate notation would be $\boldsymbol{D}_X$ and $\boldsymbol{D}_G$, but we trust that the point is clear without going through the notational motions.

[11] For an alternative way of using the Waltonian framework to understand the applicability of mathematics see Leng's (2010).

are assumed to be operable in the model. In Newton's model, for instance, the two bodies are assumed to be governed by Newton's equation of motion, which is the mathematical principle of generation that is used to find that it is fictional in the model that planets move in elliptical orbits. These principles are independent of D_X and can be changed. This happened, for instance, when, without changing D_X, Newton's equation was replaced by Schrödinger's equation to generate secondary truths about the two-body system.

How does the fact that D_G and D_X can themselves contain mathematical language fit into the extensionalisation account of the applicability of mathematics? There is no conflict here, and the explicit occurrence of mathematical language in model descriptions in no way invalidates the account. In fact, the two converge to the same point from opposite directions. Rather than first giving a physical description and then extensionalise it, we may just as well first present a mathematical structure and then find concrete terms that are the correct "interpretation" of the mathematical terms in the sense that the mathematical expressions we started with turn out to be the correct extensionalisation of whatever concrete terms they are attached to. In fact, this is what Weisberg does in his construal (which we discussed at the end of the previous section). What matters, as far as the application of mathematics is concerned, is not in which order these steps are carried out, but that the package we reach at the end contains both kinds of description (i.e. of D_S and D_S).

9.3 Limit Keys

Keys are an essential ingredient of the DEKI account. So far we have seen a few relatively straightforward keys, which were based on colour coding (litmus paper, Mandelbrot sets), scaling relations (maps, scale models), and control mechanisms (Phillips–Newlyn machine).

Keys will vary from discipline to discipline, and it is doubtful that there will ever be a complete list of keys. This should not stop us, however, from studying specific keys in particular disciplines with the aim of understanding how exactly these keys work. Indeed, it should encourage us to do so. In this section we want to make a start and discuss in some detail a kind of key that is important in physics and in engineering: limit keys. Limit keys respond to the fact that models often exemplify "extremal" features. Model-planes are frictionless; model-gases have an infinite number of molecules; model-planets are perfect geometrical spheres; and so on. Imputing such features unaltered to targets results in obvious falsities because real systems aren't frictionless, don't have an infinity of constituents, and don't have prefect geometrical shapes. This raises the question: what do models exemplifying such features tell us about target systems that don't, and never will, have such features? Limit keys provide an answer to this question by exploiting the fact that the model features can be see as resulting from taking certain features of the target to a limit.

Before we can give a definition of limit keys, we have to introduce the notion of a limit. A sequence is a succession of objects. In principle one can construct a sequence with objects of any kind, but we here restrict attention to the two most common kinds of sequences: number sequences and function sequences.

A *number sequence* is a succession of numbers linked by a rule. The list is usually indexed by an index α and the rule is given by an operation. As an example, consider the sequence $1/\alpha$ for $\alpha = 1, 2, \ldots$. We follow an often-used convention and write such sequence as f_α. In our example we have $f_\alpha = 1/\alpha$. The index in our example is a natural number. Although this choice is intuitive, nothing depends on the index being a natural number and any number can be used; in fact, we will soon see an example where α is a real number.

We can now ask how f_α behaves if α tends toward infinity. That is, we can consider the *limit* of f_α for $\alpha \to \infty$, where the symbol "∞" denotes infinity. If that limit exists and has the value L we write $\lim_{\alpha \to \infty} f_\alpha = L$. The question now is how a limit can be defined precisely and under what circumstances it exists. The standard definition of a limit is couched in terms of positive real numbers ε (where "positive" means that $\varepsilon > 0$). These numbers can be arbitrarily small but never equal to zero.[12] Then, L is the limit of the sequence f_α for $\alpha \to \infty$, i.e. $\lim_{\alpha \to \infty} f_\alpha = L$, iff for every $\varepsilon > 0$ there exists an α' in the sequence such that for all α: if $\alpha > \alpha'$, then $|f_\alpha - L| < \varepsilon$. Intuitively this means that we can keep f_α as close to L as we like by making α sufficiently large. Limits with $\alpha \to \infty$ are also referred to as *infinite limits*. If it is not possible to keep f_α as close to L as we like by making α sufficiently large, then the infinite limit does not exist. If the limit exists we say that the sequence f_α *converges* towards L. Consider again the previous example of $f_\alpha = 1/\alpha$. We can now take the limit of this sequence for $\alpha \to \infty$ and it is obvious that $\lim_{\alpha \to \infty} 1/\alpha = 0$.

Infinite limits can be taken irrespective of whether α is a natural number or a real number. If we restrict attention to cases where α is a real number, we can also ask how a sequence behaves when α tends towards a particular value α_0. For instance, we can ask how f_α behaves when α tends towards zero, or towards five. The standard definition of such a limit is couched in terms of two positive real numbers, ε and δ (where, as previously, "positive" means that both ε and δ are > 0). The definition then says that $\lim_{\alpha \to \alpha_0} f_\alpha = L$ iff for every $\varepsilon > 0$ there exists a $\delta > 0$ such that for all α: if $0 < |\alpha - \alpha_0| < \delta$, then $|f_\alpha - L| < \varepsilon$. Intuitively this means that we can keep f_α as close to L as we like by keeping α close to α_0. If this is not possible, then the limit does not exist.

It is absolutely crucial not to conflate the limit of a sequence with the value of the sequence at the limit: L and f_{α_0} are not the same mathematical objects! Consider the limit for $\alpha \to \alpha_0$. The definition of the limit requires $0 < |\alpha - \alpha_0| < \delta$ – that is, $|\alpha - \alpha_0|$ has to be strictly greater than zero – and hence in taking the limit α will never be equal to α_0. For this reason the limit L reflects how f_α behaves when α comes arbitrarily close to α_0, but *without ever reaching it*; it does *not* reflect the

---

[12] What follows is the standard definition stated in books on calculus. See, for instance, Spivak's (2006, Chap. 5).

9.3 Limit Keys

behaviour of f_α for $\alpha = \alpha_0$. The same holds for infinite limits: because α tends towards ∞ without ever reaching it, L is not the same as f_∞. To express this difference clearly we call L the *limit value* and refer to f_{α_0} (or f_∞) as the *value at the limit*.

That two values are conceptually distinct does not mean that their numerical values must be different. Assume now that both the limit value and the value at the limit exist. If they coincide, that is if $L = f_{\alpha_0}$ or $L = f_\infty$, then the limit is a *regular limit*. If the two values are different, then the limit is a *singular limit* (Butterfield 2011b, p. 1077).[13]

We will see examples of both cases later. Before discussing examples, we can now say what a limit key is. Let f_α be a feature of interest in a target system. To study the target, we construct a model in which the parameter α assumes the extremal value. Let us begin with a finite value α_0. This means that the feature exemplified by the model is f_{α_0}. Now assume (a) that the limit L of f_α exists for $\alpha \to \alpha_0$; (b) that the value f_{α_0} at the limit exists; and (c) that the limit is regular (i.e. that $L = f_{\alpha_0}$). Under these assumptions it follows that for every $\varepsilon > 0$ there exist a $\delta > 0$ such that for all α with $0 < |\alpha - \alpha_0| < \delta$ it is the case that $|f_\alpha - f_{\alpha_0}| < \varepsilon$. This can be exploited. If we consider a limit $\alpha \to \alpha_0$, the model user can infer that as long as α in the target is not more than δ away from α_0 in the model, the value of f_α in the target is no more than ε away from f_{α_0} in the model. Or, more colloquially, if the parameter α in the target is close to the model value, then the feature f_α in the target is close to the feature f_{α_0} in the model. In this way knowing the model feature gives information about the target feature. If a model user employs knowledge of limits in this way to infer from a model-feature to target-feature she uses a *limit key*. The key works by taking the exemplified feature of interest f_{α_0} in the model and converting it into a logically weaker feature of interest, namely of f_α being in the interval $\left(f_{\alpha_0} - \varepsilon, f_{\alpha_0} + \varepsilon\right)$. It is this weaker property that is imputed to the target system.

The argument is, mutatis mutandis, the same if we consider an infinite value. In this case the feature exemplified by the model is f_∞. Assume that the limit of f_α for $\alpha \to \infty$ is regular. Then the model user can infer that if α in the target is larger than a threshold α', then the value of f_α in the target is no more than ε away from f_∞ in the model.

Let us now turn to *function sequences*. The difference between a number sequence and a function sequence is that the latter is not a sequence of numbers but a sequence of functions $f_\alpha(x)$. The functions can be of any kind, but to keep things simple we consider real valued functions $f_\alpha : \mathbb{R} \to \mathbb{R}$, where '$\mathbb{R}$' denotes the real numbers. An example of such a sequence is $f_\alpha(x) = x^{-\alpha}$. A function sequence can converge toward a limit function in different ways. One of the simplest is what is known as *pointwise convergence*. The function sequence $f_\alpha(x)$ converges pointwise towards the function $L(x)$ iff for every $x \in \mathbb{R}$ the value of $f_\alpha(x)$ converges to $L(x)$. If this is the case we write $\lim_{\alpha \to \alpha_0} f_\alpha(x) = L(x)$, and mutatis mutandis for $\alpha \to \infty$.

---

[13] Although note that Butterfield recommends caution with respect to the use of the term "singular limit", given the variety of ways it is used in the literature (2011b, p. 1068). For more on this in the context of debates concerning emergence, reductionism, and supervenience, see Batterman's (2002) and Butterfield's (2011a).

We call $L(x)$ the *limit function* and $f_{\alpha_0}(x)$ the *function at the limit*. As before, the limit function and the function at the limit can, but need not, be the same. If they both exist and are identical, then the limit is regular; if not, then it is singular.

Function sequence limits can be used to reason with the model about the target in the same way as number sequence limits. If the limit is regular it follows that for all ε there exists a δ such that for all α with $0 < |\alpha - \alpha_0| < \delta$ we have $|f_\alpha(x) - f_{\alpha_0}(x)| < \varepsilon$ for all x (and, again, mutatis mutandis for $\alpha \to \infty$). This means that a model user can infer that as long as α in the target is not more than δ away from α_0 in the model, the function $f_\alpha(x)$ in the target is no more than ε away from $f_{\alpha_0}(x)$ in the model.[14] The limit key then works by taking the exemplified feature of interest $f_{\alpha_0}(x)$ in the model and converting it into a logically weaker feature of interest of $f_\alpha(x)$ being in the interval $\left(f_{\alpha_0}(x) - \varepsilon, f_{\alpha_0}(x) + \varepsilon\right)$ for all x, which is imputed to the target system.

Let us now see limit reasoning in action with two toy examples. It is always illustrative to see what happens if a method fails, and so we start with an example, based on a number sequence, in which a limit key fails. Your target system is a staircase that you want to carpet. To buy the right amount of carpet you need to know the stairs' total length. The staircase in which the steps are located has the shape of a right-angled triangle with both sides having unit length, and with the steps sitting on the hypotenuse. Further suppose that there are a large number of steps in the staircase and you somehow cannot work out their total length. You therefore resort to a model.

Let $\alpha = 1, 2, \ldots$ be the index of the number sequence. You start with a staircase with two steps and every time you progress to the next index you double the number of steps on the staircase: for $\alpha = 1$ the staircase has two steps, for $\alpha = 2$ four steps, for $\alpha = 3$ eight steps, and so on. This is illustrated in Fig. 9.2. In general the staircase with index α has 2^α steps. The dependant feature of interest, f_α, is the length of the staircase with index α; that is, f_α is the length of set of steps with 2^α steps. The number of steps seems so large to you that you model the scenario as consisting of an

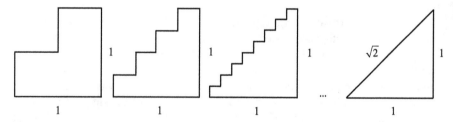

Fig. 9.2 A sequence of staircases, with the value at the limit

---

[14] Since we're using the notion of pointwise convergence, the values of δ (for each ε) can vary across different values of x.

9.3 Limit Keys

infinite number of them. A staircase with an infinite number of steps is a line, and so this idealisation results in a model, as shown by the "staircase" to the right in Fig. 9.2, where the length of the staircase is the length of the hypotenuse of a right-angled triangle whose other sides are of unitary length: $f_\infty = f_{\text{model}} = \sqrt{2}$.[15]

You of course know that the number of steps is not infinite. You think that this is not a problem because you can use a limit key. The number of steps is large, and you think that it is in fact large enough for the length of the model staircase to be close enough to the length of the real staircase for all practical purposes, in particular to buy the right amount of carpet.

This is a mistake. Looking at Fig. 9.2 it's easy to see that the total length of the staircase is two *irrespective* of the number of steps: $f_\alpha = 2$ for all $\alpha = 1, 2, \ldots$. Hence, trivially, $\lim_{\alpha \to \infty} f_\alpha = 2$. So $L \ne f_{\text{model}}$. This shows that the limit is singular, and we're now in a position to see how reasoning with a limit key breaks down when the limit is singular. From $\lim_{\alpha \to \infty} f_\alpha = 2$ we know that for every $\varepsilon > 0$ there is an α' such that: if $\alpha > \alpha'$ then $|f_\alpha - 2| < \varepsilon$, but applying the limit key would amount to mistakenly assuming that for all ε there is an α' such that if $\alpha > \alpha'$ then $\left|f_\alpha - \sqrt{2}\right| < \varepsilon$, which is false. In fact for any $\varepsilon < 2 - \sqrt{2}$ there is no α' such that if $\alpha > \alpha'$ then $\left|f_\alpha - \sqrt{2}\right| < \varepsilon$. So no matter how many stairs there are, the length of the stairs doesn't come close to the length of the hypotenuse, not even in the limit for the number of steps toward infinity! This is why a limit key doesn't work here, and you would buy the wrong length of carpet if you were to reason in this way. So by using a limit key in a case where the limit in question is singular, the model yields wrong results.

Our second example works with a function sequence and provides an illustration of a case where limit keys work. Suppose your target system is a ski-jumper and you want to know how her position on the slope changes through time. To this end you construct a model featuring a rectangular object sliding down a frictionless plane with an inclination of θ under the influence of the linear gravitational force $F = mg$, where g is the gravitational constant on the earth's surface.[16] We also assume that this is the *only* force acting on the model-ski-jumper. With some simple trigonometry we can calculate the component of the force acting on the model-ski-jumper parallel to the surface of the plane: $f^=_{\text{model}} = mg \sin(\theta)$, as shown in Fig. 9.3.

Using Newton's equation, and without loss of generality setting the original position and initial velocity to zero, we find the following position function for the model-ski-jumper:

---

[15] Thinking about this model in terms of the fiction view of models introduced above, the description D_X prescribes us to imagine a triangular object; the interpretation D_I prescribes us to interpret that object as a staircase; and the principles of generation D_G contain the mathematics required to understand that the triangle's hypotenuse is $\sqrt{2}$.

[16] Again, in terms of the fiction view, D_X is the prescription to imagine a block with rectangular profile on a perfectly straight slope with constant inclination, where both are made from materials that result in frictionless contact. D_I prescribes model users to interpret the block and the slope in terms of a skier and on a hill. D_G contains Newton's laws and the relevant mathematics required to derive the position function.

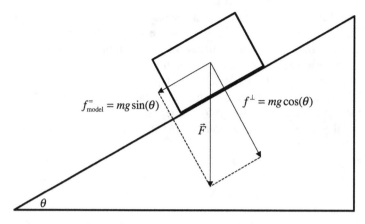

Fig. 9.3 Ski-jumper model

$$x_{\text{model}}(t) = \frac{1}{2}t^2 g \sin(\theta).$$

This function is a feature exemplified by the model. In the idiom of DEKI this function is Z_1. We know perfectly well that the real slope isn't a frictionless perfect plane, and that there are forces other than gravity acting on the skier such as air resistance and the Coriolis force. Given this, what does the model tell us about the real skier's position? To answer this question, we need a key. In keeping with the spirit of our above discussion, we understand the model as limiting case of the real situation and aim to construct a limit key.

To make a start, let us assume that the only force acting on the skier not taken into account in the model is friction, and that friction is linear. This is a strong assumption and we come back to it later – let's run with it for now to see how the reasoning works. The magnitude of the friction force acting on the skier then is proportional to the magnitude of the force perpendicular to the plane, $f^{\perp} = mg \cos(\theta)$, where the proportionality constant is the friction coefficient μ. This means that the actual position function of the skier is:

$$x_{\text{friction}}(t) = \frac{1}{2}t^2 g \left[\sin(\theta) - \mu \cos(\theta)\right].$$

Now regard μ as a freely variable parameter and notice the following relationship holds: $\lim_{\mu \to 0} x_{\text{friction}}(t) = x_{\text{model}}(t)$. To see this, and to connect it to our above definition of a limit, it suffices to notice that the relevant δ for each ε is given by $\delta(\varepsilon) = 2\varepsilon / \cos(\theta)t^2$. It's then easy to see that the definition of the limit is satisfied and that the limit is regular. This allows us to use a limit key. For all times t and for any $\varepsilon > 0$ it is the case that as long as $\mu < \delta(\varepsilon)$, it's guaranteed that $|x_{\text{friction}}(t) - x_{\text{model}}(t)| < \varepsilon$. Or in plain English: as long as the friction coefficient in the actual system is less than $\delta(\varepsilon)$, the position function in the model will differ from the actual position

9.3 Limit Keys   201

function by less than ε. Returning to the idiom of DEKI, Q_1 is the property of the position of the skier in the target being in the interval $(x_{\text{model}}(t) - \varepsilon, x_{\text{model}}(t) + \varepsilon)$ where the values of ε depends on the lower bound the model user can set on the value of the fiction coefficient μ in the target. The key connects Z_1 to Q_1, and the model imputes Q_1 to the target.

Let us now return to our assumption that the only force acting on the skier not taken into account in the model is friction, and that friction is linear. This assumption allowed us to specify the δ and the ε explicitly and prove that the limit exists. We made this assumption to illustrate how limiting reasoning works. It is, unfortunately, unrealistic in two ways. First, there are known unknowns. Even when further factors are known, it is not always possible to calculate their effect explicitly. We know that the real slope is uneven in various ways and that this unevenness has an effect on the real skier's motion. But we can't calculate this effect explicitly. Nor can we calculate the effect of the breeze of wind that just happened to blow on the skier in the moment when she starts moving. And so on. So we cannot always explicitly specify the difference between a model and the target; linear friction is a special case in that regard. Second, and worse still, there are unknown unknowns; we may not know all the factors that influence a situation. For example, the skier may be subject to forces we don't know. Knowing all relevant factors would require a God's eye perspective that mortal scientists don't have. The consequence of this is that in practice we cannot neatly quantify the differences between model and target, and we cannot rigorously prove that the model is a regular limit of a sequence that contains the real-world target.

Does this pull the rug from underneath limiting reasoning? For those who require mathematical proofs, yes. But there are rarely, if ever, mathematical proofs backing the successful *application* of a model to the world.[17] What scientists will do in this situation is to form a qualitative judgement against their background knowledge. They will take into account everything they know about forces and their effect on bodies, and they will make a qualitative estimate of the magnitude that this effect will have on the skier. This will give them an interval $(x_{\text{model}}(t) - \varepsilon^e, x_{\text{model}}(t) + \varepsilon^e)$, where the superscript "e" stands for "estimate", of which they will be willing to say that the real position of the skier will lie in that interval given everything they know about forces. This defines a property Q_1^e that they can then impute to the target.

Limits have not become obsolete. The justification for imputing Q_1^e rests on the belief that a limit exists and that the model function is only so far away from it. Let us spell this out in more detail. Meet an old friend: Laplace's Demon (Laplace 1814). The Demon knows all the forces and can write down the true position function $x_{\text{skier}}(t)$ of the skier. This function will depend on a myriad of parameters. The

---

[17] And there are good reasons to doubt that we should expect there to be such proofs. Whether or not a model is an accurate representation depends on features beyond the model: it depends on the nature of the target system in question. As such, whilst we may be able to prove by taking relevant features to the limit that *if* the target is such, *then* the model will allow us to reason successfully about the target, the antecedent of this conditional isn't the sort of thing that admits mathematical proof.

claim that scientists – mostly implicitly – rely on is that if the Demon took all of the parameters in $x_{\text{skier}}(t)$ towards their values in the model (which will be zero for the vast majority of them, given that they will be parameters corresponding to features of the skier that scientists, as limited beings, haven't included in their model), that limit would turn out to exist and to be regular. That is, they assume $\lim x_{\text{skier}}(t) = x_{\text{model}}(t)$, where we write "lim" (without subscripts) to indicate that the limit is taken for all parameters. Of course, $\lim x_{\text{skier}}(t) = x_{\text{model}}(t)$ is not provable, not least because human scientists, lacking the powers of the Demon, don't have access to $x_{\text{skier}}(t)$. It is a transcendental assumption in the sense that it is necessary in order to apply the model (based on the limit key) even though the assumption itself cannot be proven. But it is an assumption that scientists *must* make if they are to assume that the model is informative about the target. If the limit does not exist, or if it is irregular, then there is no reason to assume that the target behaves like the model, even if the model's parameter values are close to the target's parameter values.

Carpets and ski jumpers are toy examples. But the same inferential patterns are at work in "real" applications. Consider again the Newtonian model of a planet's orbit. As we have seen in the Introduction, the model involves imagining two spheres, each with a homogenous mass distribution. One of the spheres is much more massive than the other, and the only force acting in the system is the gravitational force between the sphere. Combining these assumptions with Newton's second law gives an equation of motion for the planet in the model. Of course this equation of motion isn't the exact equation of motion governing the actual planet: even supposing that Newtonian mechanics were correct, the actual force that determines how a planet moves includes forces beyond its gravitational interactions with the sun. So we have an exemplified feature of a model, $x_{\text{model}}(t)$, that is a solution to the differential equation, $\ddot{\vec{x}} = -Gm_s \vec{x}/|\vec{x}|^3$, which we know doesn't match any actual feature of the target. What, then, does the model tell us about the movement of real planets? The answer, we submit, is provided to us by a limit key. We should think of the actual trajectory $x_{\text{planet}}(t)$, available to the Demon but not to us, as being such that if the Demon took the parameters in $x_{\text{planet}}(t)$ to limits corresponding to their value in the model – again, presumably most of them will be taken to zero given that they don't appear in $x_{\text{model}}(t)$ – then the Demon would find that $\lim x_{\text{planet}}(t) = x_{\text{model}}(t)$.

This kind of reasoning has been incredibly successful throughout the history of physics, and indeed engineering. From planetary motion to rocket launches it has worked successfully in countless applications. This lends credibility to the use of limit keys in mechanics, and makes scientists confident that limit keys will also work in future applications. It is important to realise, however, that inductive support for limit reasoning does not "prove" the method right. In fact, scientists have worried about these limits time and again, and uncovering the scope of limit reasoning has been a scientific endeavour in its own right.

As an example consider Poincaré's study of the role of initial conditions. Among the parameters that $x_{\text{planet}}(t)$ contains are the planet's position and momentum at certain initial time t_0. This is because Newton's equation of motion tells us where a planet is at a later time $t > t_0$ only if we specify the planet's position and momentum at some initial time. This specification is the planet's *initial condition*. In practice scientists can only ever specify an approximate initial condition because it's impossible to measure the condition with absolute precision. Limit reasoning then says that if the initial condition of the planet in the model is sufficiently close to the initial condition of the real planet, then the trajectory of the planet in the model is sufficiently close to the real planet's trajectory. For centuries this assumption was taken for granted, until Poincaré showed that the assumption was not true in general. Poincaré studied what is now known as the three-body-system, which is exactly like the Newtonian model except that it has a third sphere in it. If you want an interpretation, you can think of these three spheres as the sun, the earth, and the moon. What Poincaré found was that the three-body-system exhibits what is now known as *sensitive dependence on initial conditions*: even if two initial conditions are arbitrarily close, their trajectories can diverge. This is now known as *chaos*.[18] This means that the limit does not exist and hence the model cannot be equipped with a limit key.

This has far-reaching consequences. Specifically, it means that Newton's model cannot be equipped with a limit key and expected to provide true results, at least not universally and unrestrictedly. What exactly the restrictions are is a question discussed in the discipline of chaos theory. The details are beyond the scope of this book, but one of the crucial results is that in contexts like the ones that Poincaré considered, limit key imputations can be expected to yield correct results only for a finite time span, and not for all times. So chaos theory tells us that the transcendental assumption is only justified for finite times.

Questions about limits go beyond initial conditions. What happens if the dynamics of the target system is different from the dynamics of the model in certain respects? This question prompted a study of what is known as *structural stability*, which continues to date.[19] So the study of the limitations of limits is not only a philosophically interesting issue; it is also a field of active scientific research.

Coming back to the larger theme of the nature of keys in general, we are in no way claiming that limit keys exhaust the kinds of keys used in epistemic representation. The keys we have discussed previously, associated with colour coding, scaling relations and tolerance thresholds, aren't analysed in terms of limits. But by discussing limit keys in detail we hope to have illustrated that it is clear how analysing instances of model-based science in terms of DEKI sheds light on the practices in question, and related philosophical questions that arise from it. This acts as a "proof-of-concept". Our hope, then, is that more detailed investigations into different kinds of keys will further expand our understanding of other areas of model-based science. But that is a task for another day.

---

[18] For a discussion of Poincaré's discovery of sensitive dependence on initial conditions see Parker's (1998), and for a discussion of the implications of chaos for predictability see Werndl's (2009). For accessible introductions to chaos see, for instance, L.A. Smith's (2007) and P. Smith's (1998). For an advanced discussion see, for instance, Lichtenberg and Liebermann's (1992).

[19] For a technical discussion of results see Pilyugin's (1991). Frigg et al. (2014) provide an accessible introduction and a discussion of philosophical consequences.

9.4 Beyond Models

As we have seen in Sect. 8.5, the programme of analysing epistemic representation in terms of representation-as gives a negative answer to the Scientific Representational Demarcation Problem and is thereby committed to the view that there is no difference between scientific representations and other kinds of epistemic representations as regards their representational characteristics. The programme's answer to the Taxonomic Scientific Representational Demarcation Problem is "minimal" in the sense that it only distinguishes between direct and indirect epistemic representations, and on some versions not even that. In this section and the next we reflect on the issue of taxonomic demarcation and then discuss what a rejection of scientific demarcation involves.

Let us begin with the Taxonomic Representational Demarcation Problem, the question whether there are different types of representations. A specific version of this problem is the question whether, if at all, models differ from other forms of representations. The examples we discussed in Chap. 8 motive the claim that there is no difference between models and visual representations. After a short immersion in a solution, a strip of litmus paper exemplifies a certain shade of red, and, via a key that converts a colour spectrum into levels of acidity, ascribes a pH value of 3.5 to the solution. In the representation of the Mandelbrot set in Sect. 8.6, Fig. 8.5, a key is used that translates colour into divergence speed. Diagrams, graphs, and other visual representations can be analysed along the same lines.

A harder question is whether scientific theories can also be analysed within the DEKI framework. An immediate problem in dealing with this question is that it's not clear what a theory is, and conflicting analyses have been offered. Specifically, as we have gestured at in previous chapters, the relationship between theories and models has been discussed controversially (see Portides' (2017) for a review). Relevant to our current purposes are three positions. First, the so-called semantic view of theories takes a scientific theory to be a collection of models (see, for instance, Giere's (1988) and van Fraassen's (1980) for statements of the view). Second, the models as mediators view, associated primarily with Morgan and Morrison's (1999) edited collection, takes scientific models to mediate between theories and the world with models playing the primary representational role.[20] Third, the now so-called syntactic view of theories identifies theories with a linguistic description of their targets in a formal language, usually with the assumption that the descriptions are to be understood literally and that models play no representational role at all. For classical statements of the syntactic view see Nagel's (1961, Chap. 5) and Hempel's (1970). For a discussion of the role of models in syntactic view see Braithwaite's (1962).

Depending on which of these positions one thinks is the correct one to understand the relationship between theories and models, different answers to the Taxonomic Representational Demarcation Problem emerge. If one identifies

---

[20] See also Cartwright et al. (1995) and Suárez and Cartwright's (2008).

9.4 Beyond Models

theories with collections of models, the DEKI account of how models represent automatically entails that it applies to how scientific theories represent. If one subscribes to the models as mediators view then scientific theories are not, strictly speaking, representational by themselves; theories are tools for model-construction and it's models that represent. On this view, then, there is a demarcation between models and theories, but this demarcation is inconsequential in the current context because theories are not seen as representational and the representational function of models is captured by DEKI.

The more significant possibility for demarcation comes if one subscribes to the third option, that is the syntactic view. Here the discussion of direct vs. indirect epistemic representation in Sect. 3.1 becomes relevant again. If theories are identified as linguistic representations that provide a literal description of their subject matter then they are direct epistemic representations, and thus don't represent in a manner that DEKI is designed to capture. Thus understood, theories have to be distinguished from indirect epistemic representations like graphs, charts, and models. As we have previously seen, Weisberg thinks that direct epistemic representation doesn't fall under the auspices of the similarity account. It seems to us that the same would have to be said about the DEKI account. The final jury on this is still out and much will depend on one's philosophy of language, but, at least prima facie, declarative sentences like "the cat is on the mat" or "a coral atoll is a ring-shaped coral reef that encircles a lagoon" would not seem to relate to their subject matters by exemplification, keying-up and imputation. Hence, those who accept that there are direct epistemic representations, either because they subscribe to the syntactic view of theories or for some other reason, will restrict the scope of the DEKI account to indirect epistemic representations and accept that the workings of direct epistemic representation would have to be explained in another account, presumably to be provided by the philosophy of language.

Let us now turn to the Scientific Representational Demarcation Problem, the question how, if at all, scientific epistemic representations differ from other kinds of epistemic representations. Again, DEKI's answer is that they don't. This is hardly surprising given that essential components of the account were motivated by maps and pictures. But DEKI is not limited to "close cousins" of scientific representations, and it can be put to work in artistic representation too. In fact, all its key components play important roles in understanding works of art.

Let us begin with interpretations, which turn "mere" objects into Z-representations. The fact that we readily recognise Edgar Degas' *The Rehearsal of the Ballet Onstage* ('*Rehearsal*' for short) as a ballet-representation may mask the fact that this recognition is the product of an interpretation. In Sect. 1.1 we discussed the warning that Maurice Denis issued to his fellow painters: before being a portrayal of an object or a narrative, a picture is a flat canvas covered with pigments. Per se, a painting is a welter of lines and dots, a bounded collection of curves, shapes, and colours. Before we interpret Degas' canvas in certain way, it is neither a representation-of something, nor a something-representation. Assume that we make a temperature measurement at each point of a surface (for instance the bonnet of a car) and use a colour-coding similar to the one used for the Mandelbrot set to record the outcomes in the form of a plot. Further assume that it so happens that the temperature distribution is such that the resulting temperature plot is visually indistinguishable from *Rehearsal*. Would we say that this plot is a ballet-representation? No. A coloured surface that looks like *Rehearsal* is a ballet-representation only under an interpretation that takes the colours of the surface to be representations of a visual experience we have when seeing ballet dancers.[21]

Emphasising the importance of an interpretation in understanding a visual pattern is more than just an academic point. Much confusion can be avoided by bearing in mind that visual patterns are not "natural" depictions of something just because they look like something, where "natural" is taken to mean that there is some objective relation between the depiction and the depicted that does not depend in any essential way on the role of onlookers. This point is brought home by the case of Putnam's ant, which traces a line through the sand that ends up looking like Churchill (Sect. 3.2). Has the ant produced a picture of Churchill? No it hasn't. Although the visual similarity between the trace in the sand and the British politician *can* form the basis of such an interpretation (an onlooker could interpret the shape of the trace as the shape of Churchill's face with a cigar in his mouth for example), it needn't. And without an onlooker there is no interpretation to begin with and the trace is not a Z-representation of any kind.

The importance of an interpretation is highlighted by considering cases where the "obvious" or "natural" understanding of an image is in fact not the correct one. Elkins discusses striking cases of such images. One of his examples is a widely reproduced Hubble Space Telescope image of young stars in the Eagle Nebula (Elkins 2007, pp. 10–12). We see an image that looks like an under-water photograph of a rock formation that is covered with a thin layer of brownish seaweed. The unsuspecting onlooker is seduced into thinking that young stars in the Eagle Nebula look like seaweed-covered rock formations, and part of the popularity of such images derives from the seemingly easy visual access that they provide to astronomical phenomena. But, as Elkins points out, this reading of the image is profoundly mistaken. The image is a fusion of thirty-two individual images taken with four different cameras. These images were cleaned, stitched together, and given false colours. The colours that appear to represent an ordinary visual impression in fact are a coding for physical properties of the objects (blue, for instance, stands for the emission of doubly ionised oxygen). Unsuspecting onlookers unaware of all this will radically misinterpret the image.

In some cases, pictures that are prima facie misleading raise interesting questions. Mandelbrot (1982) presents an impressive collection of images that look like depictions of mountains and planets even though they are the result of mathematical algorithms and colour codings of the kind described in Sect. 8.6. Barnsley (1993) produced a welter of images of the same kind that look like ferns. These, and similar achievements, were hailed as the discovery of the "fractal geometry of nature" (as

---

[21] This example is also discussed in our (2017c). Explaining how this kind of interpretation works is no easy feat. See Kulvicki's (2006b) for a useful review of the options discussed in the philosophy of art.

9.4 Beyond Models

Mandelbrot called it). It is surely remarkable that pictures that look like ferns can be produced by mathematical algorithms plus a colour coding scheme, but announcement of the discovery of the fractal geometry of nature may well be premature. Per se these images tell us more about an onlooker's interpretation than about nature itself. Filling the gap between appearance and an underlying mechanism has become the subject matter of the field of research known as fractal growth theory, which aims to show that the equations generating the images can be seen as representations of real physical or biological processes, and that the shapes seen in the computer-generated images therefore reflect natural process. If true, that's a significant discovery, and one that goes way beyond the superficial observation that a computer plot, when seen through a visual-image-interpretation, looks like a fern or a planet.

DEKI also has the means to explain the working of symbolic art. Frans Pourbus the Younger's painting of Anne of Austria is, in our parlance, a Princess-with-dog-representation. The painting is also a representation-of Princess Anne, because it denotes the princess. But it is not a representation-of her dog (even if she had one); the part of the painting showing a dog does not denote anything because the painting doesn't function like a portrait of a royal couple where half of the painting denotes the queen and the other half the king. But the dog is an important part of the picture and can't be dismissed as a mere ornament. The dog is exemplified. Under the conventions used at the time the dog was a symbol for fidelity, and so the painting should be read as coming with a key associating a dog with fidelity (in much the same way in which litmus paper comes with a key associating the colour red with acidity). The painting then imputes the thus keyed-up property to the princess and represents her as faithful.

Similar moves can be brought to bear on modern art, and on art that isn't pictorial. A flame is moving along a fuse. It reaches a tyre, which starts rolling down a slope. It reaches the ground and moves horizontally for a short while before it starts climbing a tilted balance, its speed being just sufficient to pass the midpoint. This tips the balance to the other side and the tyre rolls down again. After having gone up and down another smaller balance it hits a board that is tied to a ladder. The ladder falls, hitting another board, which kicks the tyre in the direction of an oil barrel on top of which there is small trolley with a burning candle. The trolley starts moving and soon gets stuck under a metal grid with sparklers, which catch fire. This lights another fuse, setting off a small firework. A spark of the firework ignites a puddle of oil, and so on.

This is the opening sequence of the 1987 film *The Way Things Go* by Swiss artists Fischli & Weiss.[22] In the 29 minute long film we see a seemingly endless sequence of events involving physical objects such as tires, ladders, oil barrels, shoes, and soap. The events are carefully arranged and subtly calibrated. They unfold according to exceptionless laws and yet there is an element of surprise in them. The sequence of events fascinates and even creates sense of suspense about

---

[22] A sequence of the movie can be seen at https://www.youtube.com/watch?v=GXrRC3pfLnE.

what's next (a reviewer for *The Independent* enthusiastically reported that watching *The Way Things Go* was like watching a Hitchcock movie). Yet there is no purpose, no cause, no finality, and no meaning to either the events themselves, or to their progression. What happens is aimless and eventually pointless.

Yet the movie is not just a piece of somewhat unusual entertainment. The title of the movie has an unmistakably existential ring to it and can be seen as making reference to the fate of human ambition, the purpose of social struggle, and the search for meaning in life.[23] The sequence of events can be interpreted existentially which makes the movie a human-condition-representation. Under this interpretation it exemplifies the feature of consisting of a sequence of carefully calibrated but ultimately aimless events. The movie can be taken to denote human lives, and impute this feature to these lives. In this way it represents the conditio humana as a sequence of carefully calibrated but ultimately aimless events.[24]

So we see that the Phillips–Newlyn machine, a scientific model, and the artwork *The Way Things Go,* have more in common that one would have expected: they both *represent* their respective targets *as* thus or so in the sense of DEKI. The Phillips–Newlyn machine represents the Guatemalan economy *as* an open IS-LM economy and *The Way Things Go* represents life *as* a sequence of carefully calibrated but ultimately aimless events.

DEKI also can be brought to bear on works of literature. The objects of fictional stories can be interpreted in the same way in which objects in scientific models, both physical and fictional, are interpreted. Phillips and Newlyn interpreted the hydraulic features of their machine as economic features, and Newton interpreted his two perfect spheres as a planet and the sun. Some works of literature can be seen as working in the same way. George Orwell's *Animal Farm* tells the story of a farm that is run by the animals. But the novel is not a manifesto for the self-governance of non-humans or a demonstration of the intelligence of pigs. The novel is an allegorical denunciation of Soviet-style communism as an exploitative reign of terror. The pigs are to be interpreted as the party functionaries and other animals – horses, chicken, sheep, and so on – as other segments of society; the happenings on the farm are to be interpreted as political events. Thus interpreted, *Animal Farm* is Soviet-communism-representation. As such it need not be a representation-of any particular country or party apparatus. But in a letter to a friend Orwell described the novel as a tale against Stalin, indicating that the novel denotes Soviet Russia during the first half of the twentieth century, and a number of characters in the novel denote concrete historical figures: the pig called Napoleon denotes Stalin, Snow Ball denotes Trotsky, Squealer denotes Molotov, etc. The plot exemplifies a number of features like power being built on a cult of personality while loyalty and hard work aren't rewarded, decisions being arbitrary, and innocent creatures being sacrificed mercilessly in power games of a ruthless and selfish elite. All these are imputed

---

[23] This is clearer in the original German title *Der Lauf Der Dinge*.

[24] We briefly mention alternative interpretations in the next section.

9.4 Beyond Models

(with an identity key) to Stalin and his entourage, thus providing a piercing criticism of the phoney pretensions of communism.[25]

Voltaire's *Candide: or, Optimism* tells the story of a young man, Candide, who adheres to the teachings of Professor Pangloss and believes that everything in the world is for the best. But when he starts travelling the world, experiencing hardship, disaster, and suffering, he becomes disillusioned with Pangloss' doctrine, which he comes to see as fundamentally at odds with how things are. On the face of it, the book is a story about the adventures of a good-hearted but naïve traveller, and the story unmasks Pangloss' optimism as a doctrine that is at odds with the course events in the world. But we miss an important point if we stop here. Voltaire wrote the book as a response to Leibniz's doctrine that we live in the best of all possible worlds, created by a benevolent and omniscient God. In fact, Pangloss is a parody of Leibniz and so we should read Pangloss as denoting Leibniz. The story exemplifies there being an unbridgeable gap between optimist teachings and real-world events, denouncing the optimist doctrine as a piece of bogus philosophy. These features are imputed to Leibniz's philosophy (again with an identity key), and Leibniz himself is portrayed as a promulgator of a delusional and ultimately dishonest vision of the world.

These two examples aren't handpicked exceptions. Satirical and allegorical works can generally be interpreted in the same manner as the above, and so can fables and parables. Realist fiction also fits the mould (as we will see in the next section), and so do historical and biographical novels. Notice then, that if this is the correct way of thinking about how various works of literature can inform us about the world, then our discussion has direct implications for the work in the philosophy of art concerning the cognitive value of art. There the debate concerns whether or not works of art, including works of fiction, have epistemic value and if they do, whether or not this contributes to the works' aesthetic value.[26] "Cognitivists" are committed to both (i) the idea that art, at least in part, has epistemic value because we learn from it, and (ii) that art's epistemic function positively contributes to its aesthetic value. "Anti-cognitivists" deny one, or both, of these claims. Anti-cognitivist arguments usually proceed from the observation that science is the paradigmatic provider of cognitive value, and since, so the argument goes, art doesn't work like science, if it has epistemic value at all, then it must be of a fundamentally different kind to that which we arrive at via scientific investigations (Stolnitz's (1992) is a classic example). Our rejection of the Scientific Representational

---

[25] An alternative analysis would take the story at face value and see the plot as an animal-farm-representation. The conversion of animal-farm-features into Soviet-communism-features would then be put into the key. We are not adjudicating between these options here. In our view it is a strength of the framework that it has the flexibility to accommodate different analyses of a work of literature.

[26] Both Gaut (2003) and Thomson-Jones (2005) characterise the debate about art in this way. Gibson (2008) prefers a different characterization according to which the question is whether the artwork itself, qua artwork, contains cognitive content. The difference is not important for our current project.

Demarcation Problem suggests another way of proceeding: rather we should investigate how DEKI can be brought to bear in understanding the epistemic value of art.

9.5 Difference in Sameness

So far we stressed the parallels between representation in art and science, and argued that both can be accommodated within the DEKI framework. We do, however, not mean to suggest that representation in art and science is the same in all respects. There are important differences. But these, we claim, are differences of degree rather than kind. In this section we want to highlight some of these differences. We concentrate on a few selected issues: the role of targets, the flexibility of interpretation, and the importance of rhetoric and style. To keep the discussion manageable we restrict attention to literature; similar points could be made about other art forms.

A fundamental objection to the project of drawing parallels between representation in art and science is that artistic representations have no well-defined target. Writing specifically about literary fiction, Currie notes that "[w]e have no more than the vague suggestion that fictions sometimes shed light on aspects of human thought, feeling, decision, and action" (2016, p. 304). Since we don't find real-life analogues of, say, Natasha and Pierre (in Tolstoy's *War and Peace*) we cannot compare the novel and the world, which pulls the rug from underneath the project of likening representation in art and science, because such a comparison is a defining feature of scientific modelling.

The contrast between scientific models and literary fiction is rather less stark. Neither do all models have unproblematic real-world targets; nor do all works of literature lack targets. Turning to models first, we note that many models have parts or aspects that lack real-life analogues. Models involving the ether, phlogiston, Ptolemaic epicycles, and Lamarckian inheritance of acquired characteristics are cases in point. Furthermore, as we have noted on various occasions throughout this book, some scientific models lack targets entirely. Some of these examples are failed models in the history of science. Others are targetless "by design": models of four-sex reproduction in population dynamics (Weisberg 2013), the φ^4-model in quantum field theory (Hartmann 1995), the Kac-ring model in statistical mechanics (Werndl and Frigg 2015), Norton's Dome in classical mechanics (Norton 2003, 2008), and the baker's map in chaos theory (Frigg et al. 2016) are all models without targets. Some targetless models are special in the sense that they serve the purpose of guiding us in how to construct a system (e.g. an architectural model) and thus at their moment of use lack a target system.[27] These models aren't failures. They were

---

[27] A. Currie (2017) argues that the fiction view of models has difficulty accounting for such models, given that there is no target system whose features we can compare with the model features, something which he assumes is necessary for representation. In response to this objection we note that (i) the DEKI account doesn't require feature comparisons (cf. footnote 3) and, more importantly,

known all along not to have targets, and they were constructed for purposes other than the exploration of a particular target.[28]

Turing to literature, we note that not all works of literature lack targets. As we have seen previously, satirical novels like *Animal Farm* and *Candide: or, Optimism* can have clearly specified targets. Biographical novels like Vargas Llosa's *Aunt Julia and the Scriptwriter* are tales about real-world characters. Works in the tradition of social realism such as Émile Zola's *Germinal* and Charles Dickens' *Oliver Twist* offer piercing commentary on social reality and fierce criticism of poverty. Erich Maria Remarque's *All Quiet on the Western Front* and Kurt Vonnegut's *Slaughterhouse-Five* are passionate denunciations of the horrors of World Wars I and II, respectively. And so on.

One may argue that the horrors of war or Stalin's cult of personality are too broad and unspecific to serve as targets. Maybe they are, and there is a discussion to be had about what counts as a target system and how it is delineated. But it pays to note that also in scientific contexts not all target systems are precisely circumscribed. Economic models represent general phenomena such as unemployment, inflation, business cycles, and exposure to risk; ecological models are about general processes such as population growth and predator-prey dynamics; physical models represent the approach to equilibrium; sociological models portray social exclusion; and political scientists have models of conflict resolution. None of these are specific in the sense that they denote a particular, spatiotemporally circumscribed target like our solar system. Hence, if there is difference in specificity between the targets of literary fiction and scientific models, then the difference seems to be one of degree rather than kind. The dimensions along which comparisons could be made is largely uncharted territory.

The grain of truth in Currie's observation is that not all novels have even a vague target. Franz Kafka's *The Castle* or Fyodor Dostoevsky's *Crime and Punishment* are not about anything in particular, at least not in any obvious way. They are not about World War II or poverty. This does not mean, however, that readers cannot take these novels to be about specific things. The plaintiff trying to manoeuvre her way through the endless and often uncooperative positions of a contorted legal system may interpret *The Castle* as being about her legal nightmare; and the remorseful criminal can recognise himself in Raskolnikov. The choice of a target in such cases is ad hoc, and a myriad of other targets are equally possible. Readers are free to choose targets, and when they do so they can use the novel to generate insights about their chosen target. It seems to be correct to say that this kind of underdetermination of targets is more common in literature than in science, but at the same time it should be acknowledged that the phenomenon is not unheard of in science either. The harmonic oscillator is the physicist's favourite workhorse and almost

---

(ii) the DEKI account does not assume that a model needs a target to be classified as a model (cf. Sect. 8.2).

[28] Indeed, it has been emphasised variously in the debate about models that models perform a number of functions other than representation. See Knuuttila's (2005, 2011), Bokulich's (2009), Peschard's (2011), Kennedy's (2012), and Luczak's (2017) for discussions.

anything from the atoms in a wall to insulin receptors has at some point or other been modelled as a harmonic oscillator.

A point where the difference between science and art is more pronounced is the flexibility of interpretation (in the sense of DEKI). In scientific cases the Z is usually fixed by the context and the interpretation is highly regimented. Someone who doesn't interpret the large sphere as the sun simply doesn't understand the Newtonian model. In literature there often is, and indeed should be, more flexibility. How much flexibility there is depends on the context and the genre.[29] There is little flexibility in interpreting *Animal Farm* while there are (almost) no limits to an interpretation of *The Castle*. Fischli&Weiss' film also lends itself to different interpretations. We interpreted it as a conditio-humana-representation. Someone else might emphasise the borderline functionality of the arrangement and its constant risk of failure, and therefore see it as risk-representation. Feminists might point to the masculine character of the materials and see the design of the setup as a manifestation of the male preoccupation with mechanical processes; for them *The Way Things Go* could be a gender-ideology-representation. And so on. In artistic contexts the interpretation is often deliberately left open, and coming up with an interesting interpretation is a creative act in its own right. Such freedom is foreign to science, where interpretations are regimented and controlled.

A last point we want to consider is the importance of rhetoric and style in the presentation of a model or a work of literature. Language and rhetoric are crucial aspects of a work of literature. We admire great authors not only for the inventiveness of their plots, but also (and sometimes even more so) for their use of language, the elegance of their expressions, and the fluency of their diction. This importance of language and rhetoric, opponents of a parallelism between modelling and fiction point out, is entirely foreign to science. Currie submits that "[m]odels are not dependent for their value in learning on any particular formulation" (2016, p. 305), while formulations are crucial in literature. A recounting of the plot of *Hundred Years of Solitude* in the language of a seven-year-old is not the work of art that Gabriel García Márquez created, but every reformulation of Newton's model is still Newton's model.

There is no question that language and rhetoric play a different role in literature than they play in scientific models, but that does not mean that models are completely independent of their formulation. Everybody who has ever spent time solving differential equations will know that the choice of the right coordinate system for the description of the situation is crucial. In a recent paper discussing models (understood as imaginary entities) Vorms (2011) points out that what she calls the "format of a representation" is crucial to the inferences scientists can draw from a model, and a similar observation has been made by Knuuttila (2017). The very same model, when presented in different ways can yield different predictions and offer different explanations. Formulation matters. So, once again, the difference is one of degree and detail rather than kind.

---

[29] See Eco's (1992, 1994) for discussions about the limits of interpretation.

9.5 Difference in Sameness

The DEKI account of representation, building on Goodman and Elgin's notion of representation-as, highlights the commonalities between scientific and artistic representation. By understanding how each of DEKI's conditions are met we come to understand how a hydraulic system like the Phillips-Newlyn-machine can represent the Guatemalan economy as an open IS-LM economy, and how a cleverly calibrated sequence of rolling tires and burning barrels can represent the conditio humana as a string of aimless events. The account explains, in general, how an object X represents a target T as Z. This is not to say that representation-as works in exactly the same way in science as in art (or even to say that it works in exactly the same way across the sciences or across the entire field of art). DEKI's conditions are stated at the appropriate level of abstraction so that they can be met in different ways in different cases, as we have discussed. But the differences that emerge in different instances depend on how the same conditions are met.

We conclude by re-emphasising that our analysis is aimed at cases of scientific and artistic *representation*. We don't want to claim that all scientific models, let alone works of art, play representational roles. But where they do, we hope that analysing them through the lens of DEKI will help us understand how they work.

By now we hope to have shown how the DEKI account of epistemic representation works, and how by utilising it we can arrive at a deeper understanding of particular instances of such representation. Moreover, we hope that our analysis sheds light on how scientific representation relates to other kinds of epistemic representation, and thereby helps us understand how we use epistemic representation the way we do, in many different areas of life. It should be clear that the DEKI account provides a programme for future research. Different kinds of keys should be investigated, and so should how the account can contribute to the debates mentioned in the introduction: if models represent in the way that we say they do, then how should we think about the debates surrounding realism, explanation, and understanding? Finally, then, even if one is not convinced by the details of our account, we hope that the conceptual space in which we have situated questions of scientific representation – the problems and conditions introduced in Chap. 1 which have structured our discussion throughout this book – will prove useful in framing further debates concerning representation in science, and, hopefully, beyond.

Bibliography

Abell, C. (2009). Canny resemblance. *Philosophical Review, 118*(2), 183–223.
Achinstein, P. (1968). *Concepts of science: A philosophical analysis*. Baltimore: Johns Hopkins Press.
Adams, E. W. (1959). The foundations of rigid body mechanics and the derivation of its laws from those of particle mechanics. In L. Henkin, P. Suppes, & A. Tarski (Eds.), *The axiomatic method: with special reference to geometry and physics* (pp. 250–265). Amsterdam: North-Holland.
Ainsworth, P. (2009). Newman's objection. *The British Journal for the Philosophy of Science, 60*(1), 135–171.
Aldana, E. (2011). The MONIAC: Bill Phillips's machine. *Economia Politica, XXVIII*(1), 167–170.
Ambrosio, C. (2014). Iconic representations and representative practices. *International Studies in the Philosophy of Science, 28*(3), 255–275.
Ankeny, R. A., & Leonelli, S. (2011). What's so special about model organisms? *Studies in History and Philosophy of Science, 42*(2), 313–323.
Anscombe, G. E. M. (2000). *Intention* (2nd ed.). Cambridge, MA: Harvard University Press.
Anscombe, G. E. M. (2005). Practical Inference. In M. Geach, & L. Gormally (Eds.), *Human life, action and ethics: essays by G.E.M. Anscombe*. Exeter: Imprint Academic.
Apostel, L. (1961). Towards the formal study of models in the non-formal sciences. In H. Freudenthal (Ed.), *The concept and the role of the model in mathematics and natural and social sciences* (pp. 1–37). Dordrecht: Reidel.
Argyris, J. H., Faust, G., & Haase, M. (1994). *An exploration of chaos. An introduction for natural scientists and engineers*. Amsterdam: North-Holland.
Armstrong, D. M. (1989). *Universals: an opinionated introduction*. London: Westview Press.
Aronson, J. L., Harré, R., & Cornell Way, E. (1995). *Realism rescued: how scientific progress is possible*. Chicago: Open Court.
Bailer-Jones, D. M. (1999). Tracing the development of models in the philosophy of science. In L. Magnani, N. J. Nersessian, & P. Thagard (Eds.), *Model-based reasoning in scientific discovery* (pp. 23–40). New York, NY: Kluwer Academic/Plenum Publishers.
Bailer-Jones, D. M. (2002a). Models, metaphors and analogies. In P. Machamer, & M. Silberstein (Eds.), *The Blackwell guide to the philosophy of science* (pp. 108–127). Oxford: Blackwell.
Bailer-Jones, D. M. (2002b). Scientists' thoughts on scientific models. *Perspectives on Science, 10*(3), 275–301.
Bailer-Jones, D. M. (2003). When scientific models represent. *International Studies in the Philosophy of Science, 17*, 59–74.
Bailer-Jones, D. M. (2009). *Scientific models in philosophy of science*. Pittsburgh: Pittsburgh University Press.

Balaguer, M. (2018). Fictionalism in the philosophy of mathematics. In E. N. Zalta (Ed.), *The Stanford Encyclopedia of Philosophy*. URL = <https://plato.stanford.edu/archives/fall2018/entries/fictionalism-mathematics/>.

Balzer, W., Moulines, C. U., & Sneed, J. D. (1987). *An Architectonic for Science. The Structuralist Program*. Dordrecht: D. Reidel.

Barberousse, A., & Ludwig, P. (2009). Models as Fictions. In M. Suárez (Ed.), *Fictions in science: philosophical essays on modeling and idealization* (pp. 56–73). New York: Routledge.

Barnsley, M. (1993). *Fractals everywhere*. Boston: Academic Press.

Barr, N. (2000). The history of the Phillips Machine. In R. Leeson (Ed.), *A. W. H. Phillips: collected works in contemporary perspective* (pp. 89–114). Cambridge: Cambridge University Press.

Bartels, A. (2006). Defending the structural concept of represenation. *Theoria, 21*(1), 7–19.

Batterman, R. W. (2002). *The devil in the details*. New York: Oxford University Press.

Batterman, R. W. (2009). Idealization and modeling. *Synthese, 169*(3), 427–446.

Batterman, R. W., & Rice, C. (2014). Minimal model explanations. *Philosophy of Science, 81*(3), 349–376.

Begg, D., Vernasca, G., Fischer, S., & Dornbusch, R. (2014). *Economics* (11th ed.). New York: McGraw-Hill Education.

Bell, J., & Machover, M. (1977). *A course in mathematical logic*. Amsterdam: North-Holland.

Black, M. (1973). How do pictures represent? In E. Gombrich, J. Hochberg, & M. Black (Eds.), *Art, Perception, and Reality* (pp. 95–130). Baltimore and London: Johns Hopkins University Press.

Blumson, B. (2014). *Resemblence and representation. An essay in the philosophy of pictures*. Cambridge: Open Book Publishers (https://www.openbookpublishers.com/product/282).

Boesch, B. (2017). There Is a special problem of scientific representation. *Philosophy of Science, 84*(5), 970–981.

Boesch, B. (2019a). The means-end account of scientific, representational actions. *Synthese, 196*, 2305–2322.

Boesch, B. (2019b). Resolving and understanding differences between agent-based accounts of scientific representation. *Journal for General Philosophy of Science, 50*, 195–213.

Boesch, B. (2019c). Scientific representation and dissimilarity. *Synthese, Online First. DOI:*https://doi.org/10.1007/s11229-019-02417-0.

Bogen, J., & Woodward, J. (1988). Saving the phenomena. *Philosophical Review, 97*(3), 303–352.

Bokulich, A. (2008). *Reexamining the quantum-classical relation: beyond reductionism and pluralism*. Cambridge: Cambridge University Press.

Bokulich, A. (2009). Explanatory fictions. In M. Suárez (Ed.), *Fictions in science. Philosophical essays on modelling and idealization* (pp. 91–109). London and New York: Routledge.

Bokulich, A. (2011). How scientific models can explain. *Synthese, 180*(1), 33–45.

Bolinska, A. (2013). Epistemic representation, informativeness and the aim of faithful representation. *Synthese, 190*(2), 219–234.

Bolinska, A. (2016). Successful visual epistemic representation. *Studies in History and Philosophy of Science, 56*, 153–160.

Boniolo, G. (2007). *On Scientific Representations: from Kant to a new philosophy of science*. Hampshire and New York: Palgrave Macmillan.

Boolos, G. S., & Jeffrey, R. C. (1989). *Computability and logic* (3rd ed.). Cambridge: Cambridge University Press.

Brading, K., & Landry, E. (2006). Scientific structuralism: presentation and representation. *Philosophy of Science, 73*(5), 571–581.

Brandom, R. B. (1994). *Making it explicit: reasoning, representing and discursive commitment*. Cambridge, MA: Harvard University Press.

Brandom, R. B. (2000). *Articulating reasons: an introduction to inferentialism*. Cambridge, MA: Harvard University Press.

Braithwaite, R. B. (1962). Models in the empirical sciences. In E. Nagel, P. Suppes, & A. Tarski (Eds.), *Logic, methodology and philosophy of science* (pp. 224–231). Standford: Stanford University Press.

Budd, M. (1993). How pictures look. In D. Knowles, & J. Skorupski (Eds.), *Virtue and taste* (pp. 154–175). Oxford: Blackwell.
Bueno, O. (1997). Empirical adequacy: a partial structure approach. *Studies in the History and Philosophy of Science, 28*(4), 585–610.
Bueno, O. (1999). What is structural empiricism? Scientific change in an empiricist setting. *Erkenntnis, 50*(1), 59–85.
Bueno, O. (2005). Dirac and the dispensability of mathematics. *Studies in History and Philosophy of Modern Physics, 36*(3), 465–490.
Bueno, O. (2010). Models and scientific representations. In P. D. Magnus, & J. Busch (Eds.), *New waves in philosophy of science* (pp. 94–111). Hampshire: Pelgrave MacMillan.
Bueno, O., & Colyvan, M. (2011). An inferential conception of the application of mathematics. *Nous, 45*(2), 345–374.
Bueno, O., & French, S. (2011). How theories represent. *The British Journal for the Philosophy of Science, 62*(4), 857–894.
Bueno, O., French, S., & Ladyman, J. (2002). On representing the relationship between the mathematical and the empirical. *Philosophy of Science, 69*(4), 452–473.
Butterfield, J. (2011a). Emergence, reduction and Supervenience: a varied landscape. *Foundations of Physics, 41*, 920–959.
Butterfield, J. (2011b). Less is different: emergence and reduction reconciled. *Foundations of Physics, 41*, 1065–1135.
Butterfield, J. (2014). Our mathematical universe? A discussion of some themes in Max Tegmark's recent book 'Our Mathematical Universe'. *Manuscript, arXiv:1406.4348.*
Butterfield, J. (2020). On dualities and equivalences between physical theories. Forthcoming in C. Wüthrich, B. Le Bihan, & N. Huggett (Eds.), *Philosophy beyond spacetime*. Oxford: Oxford University Press.
Byerly, H. (1969). Model-structures and model-objects. *The British Journal for the Philosophy of Science, 20*(2), 135–144.
Callender, C., & Cohen, J. (2006). There is no special problem about scientific representation. *Theoria, 21*(55), 7–25.
Cartwright, N. (1983). *How the laws of physics lie*. Oxford: Oxford University Press.
Cartwright, N. (1999a). *The dappled world: a study of the boundaries of science*. Cambridge: Cambridge University Press.
Cartwright, N. (1999b). Models and the limits of theory: quantum hamiltonians and the BCS models of superconductivity. In M. Morgan, & M. Morrison (Eds.), *Models as mediators: perspectives on natural and social science* (pp. 241–281). Cambridge: Cambridge University Press.
Cartwright, N. (2010). Models: parables v fables. In R. Frigg, & M. C. Hunter (Eds.), *Beyond mimesis and convention. Representation in art and science* (pp. 19–32). Berlin and New York: Springer.
Cartwright, N., Shomar, T., & Suárez, M. (1995). The tool-box of science. In W. E. Herfel, W. Krajewski, I. Niiniluoto, & R. Wojcicki (Eds.), *Theories and models in scientific processes* (pp. 137–150), Poznan Studies in the Philosophy of Science and the Humanities, Vol. 44. Amsterdam and Atlanta, GA: Rodopi.
Caulton, A. (2015). The role of symmetry in the interpretation of physical theories. *Studies in History and Philosophy of Modern Physics, 52*, 153–162.
Chakravartty, A. (2001). The semantic or model-theoretic view of theories and scientific realism. *Synthese, 127*(3), 325–345.
Chakravartty, A. (2010a). Informational versus functional theories of scientific representation. *Synthese, 172*(2), 197–213.
Chakravartty, A. (2010b). Perspectivism, inconsistent models, and contrastive explanation. *Studies in History and Philosophy of Science, 41*(4), 405–412.
Chang, H. (2004). *Inventing temperature: measurement and scientific progress*. New York: Oxford University Press.

Chang, H. (2012). *Is Water H2O?* (Boston Studies in the Philosophy and History of Science, Vol. 293). Dordrecht: Springer.
Colyvan, M. (2013). Idealisations in normative models. *Synthese, 190*, 1337–1350.
Contessa, G. (2007). Scientific representation, interpretation, and surrogative reasoning. *Philosophy of Science, 74*(1), 48–68.
Contessa, G. (2010). Scientific models and fictional objects. *Synthese, 172*(2), 215–229.
Contessa, G. (2011). Scientific models and representation. In S. French, & J. Saatsi (Eds.), *The continuum companion to the philosophy of science* (pp. 120–137). London: Continuum Press.
Contessa, G. (2016). It ain't easy: fictionalism, deflationism, and easy arguments in ontology. *Mind, 125*(499), 763–773.
Crittenden, C. (1991). *Unreality: the metaphysics of fictional objects*. Ithaca and London: Cornell University Press.
Cummins, R. (1991). *Meaning and mental representation*. Cambridge, MA: MIT Press.
Curran, D. (2018). From performativity to representation as intervention: rethinking the 2008 financial crisis and the recent history of social science. *Journal for the Theory of Social Behaviour, September*, 1–19.
Currie, A. (2017). From models-as-fictions to models-as-tools. *Ergo, 4*(27), 759–781.
Currie, G. (1990). *The nature of fiction*. Cambridge: Cambridge University Press.
Currie, G. (2016). Models as fictions, fictions as models. *The Monist, 99*, 296–310.
Cyr, S., Ih, K.-D., & Park, S.-H. (2011). Accurate reproduction of wind-tunnel results with CFD. *SAE Mobilus, Technical Paper 2011-01-0158*.
Da Costa, N. C. A., & French, S. (1990). The model-theoretic approach to the philosophy of science. *Philosophy of Science, 57*(2), 248–265.
Da Costa, N. C. A., & French, S. (2000). Models, theories, and structures: thirty years on. *Philosophy of Science, 67 (Proceedings of the 1998 Biennial Meetings of the Philosophy of Science Association. Part II: Symposia Papers)*, S116–127.
Da Costa, N. C. A., & French, S. (2003). *Science and partial truth: a unitary approach to models and scientific reasoning*. Oxford: Oxford University Press.
Danto, A. (1981). *Transfiguration of the commonplace: a philosophy of art*. Cambridge, MA and London: Harvard University Press.
Dardashti, R., Hartmann, S., Thébault, K. P. Y., & Winsberg, E. (2019). Hawking radiation and analogue experiments: a bayesian analysis. *Studies in History and Philosophy of Modern Physics, 67*, 1–11.
Dardashti, R., Thébault, K. P. Y., & Winsberg, E. (2017). Confirmation via analogue simulation: what dumb holes could tell us about gravity. *The British Journal for the Philosophy of Science, 68*(1), 55–89.
Davies, D. (2007). Thought experiments and fictional narratives. *Croatian Journal of Philosophy, 7*(19), 29–45.
de Chadarevian, S. (2004). Models and the making of molecular biology. In S. de Chadarevian, & N. Hopwood (Eds.), *Models: the third dimension of science* (pp. 339–369). Stanford: Stanford University Press.
de Donato Rodriguez, X., & Zamora Bonilla, J. (2009). Credibility, idealisation, and model building: an inferential approach. *Erkenntnis, 70*(1), 101–118.
Decock, L., & Douven, I. (2011). Similarity after Goodman. *Review of Philosophy and Psychology, 2*(1), 61–75.
Demopoulos, W. (2003). On the rational reconstruction of our theoretical knowledge. *The British Journal for the Philosophy of Science, 54*(3), 371–403.
Denis, M. (2008). Definition of neo-traditionalism. In C. Harrison, P. Wood, & J. Gaiger (Eds.), *Art in theory 1815–1900. An anthology of changing ideas* (pp. 862–869). Oxford: Blackwell.
Dewar, N. (2019). Sophistication about symmetries. *The British Journal for the Philosophy of Science, 70*(2), 485–521.
Díez, J. (1997a). A hundred years of numbers. An historical introduction to measurement theory 1887–1990. Part I. *Studies in History and Philosophy of Science, 28*(1), 167–185.

Díez, J. (1997b). A hundred years of numbers. An historical introduction to measurement theory 1887–1990. Part II. *Studies in History and Philosophy of Science, 28*(2), 231–265.

Díez, J. (2020). An ensemble-plus-standing-for account of scientific representation: no need for (unnecessary) abstract objects. In C. Martínez-Vidal, & J. L. Falguera (Eds.), *Abstract objects. For and against* (pp. 133–149). Cham: Springer.

Donnellan, K. S. (1968). Putting Humpty Dumpty together again. *Philosophical Review, 77*(2), 203–215.

Downes, S. M. (1992). The importance of models in theorizing: a deflationary semantic view. *Philosophy of Science. Proceedings of the 1992 Biennial Meetings of the Philosophy of Science Association. Part I: Contributed Papers*, 142–153.

Downes, S. M. (2009). Models, pictures, and unified accounts of representation: lessons from aesthetics for philosophy of science. *Perspectives on Science, 17*(4), 417–428.

Downes, S. M. (2012). How much work do scientific images do? *Spontaneous Generations: A Journal for the History and Philosophy of Science, 6*(1), 115–130.

Doyle, Y., Egan, S., Graham, N., & Khalifa, K. (2019). Non-factive understanding: a statement and defense. *Journal for General Philosophy of Science, 50*, 345–365.

Ducheyne, S. (2008). Towards an ontology of scientific models. *Metaphysica, 9*(1), 119–127.

Ducheyne, S. (2012). Scientific representations as limiting cases. *Erkenntnis, 76*(1), 73–89.

Dummett, M. (1991). *Frege: philosophy of mathematics*. London: Duckworth.

Eco, U. (1992). Interpretation and overinterpretation. In S. Collini (Ed.), *Interpretation and overinterpretation: Tanner Lectures in human values*. Cambridge: Cambridge University Press.

Eco, U. (1994). *The limits of interpretation*. Bloomington and Indianapolis: Indiana University Press.

Eddon, M. (2013). Quantitative properties. *Philosophy Compass, 8*(7), 633–645.

Einstein, A. (1920/1999). *Relativity: the special and general theory*. London: Methuen.

Elgin, C. Z. (1983). *With reference to reference*. Indianapolis and Cambridge: Hackett.

Elgin, C. Z. (1996). *Considered judgement*. Princeton: Princeton University Press.

Elgin, C. Z. (2004). True enough. *Philosophical Issues, 14*(1), 113–131.

Elgin, C. Z. (2010). Telling instances. In R. Frigg, & M. C. Hunter (Eds.), *Beyond mimesis and convention: representation in art and science* (pp. 1–18). Berlin and New York: Springer

Elgin, C. Z. (2017). *True enough*. Cambridge, MA: MIT Press.

Elkins, J. (1999). *The domain of images*. Ithaca and London: Cornell University Press.

Elkins, J. (2007). *Visual practices across the university*. München: Wilhelm Fink Verlag.

Enderton, H. B. (2001). *A mathematical introduction to logic* (2nd ed.). San Diego and New York: Harcourt.

Evans, G. (1982). *The varieties of reference*. Oxford: Oxford University Press.

Fine, A. (1993). Fictionalism. *Midwest Studies in Philosophy, 18*, 1–18.

Fine, A. (1998). Fictionalism. In E. Craig (Ed.), *Routledge Encyclopedia of Philosophy. Volume 3* (pp. 667–668). London: Routledge.

Fletcher, S. C. (2020). On representational capacities, with an application to general relativity. *Foundations of Physics, 50*, 228–249.

French, S. (2000). The reasonable effectiveness of mathematics: partial structures and the application of group theory to physics. *Synthese, 125*(1/2), 103–120.

French, S. (2003). A model-theoretic account of representation (or, I don't know much about art ... but I know it involves isomorphism). *Philosophy of Science, 70*(5), 1472–1483.

French, S. (2010). Keeping quiet on the ontology of models. *Synthese, 172*(2), 231–249.

French, S. (2014). *The structure of the world. Metaphysics and representation*. Oxford: Oxford University Press.

French, S. (2017). Identity conditions, idealisations and isomorphisms: a defence of the semantic approach. *Synthese, Online First. DOI:*https://doi.org/10.1007/s11229-017-1564-z.

French, S., & Ladyman, J. (1999). Reinflating the semantic approach. *International Studies in the Philosophy of Science, 13*, 103–121.

French, S., & Saatsi, J. (2006). Realism about structure: the semantic view and nonlinguistic representations. *Philosophy of Science, 73*(5), 548–559.

French, S., & Vickers, P. (2011). Are there no things that are scientific theories? *The British Journal for the Philosophy of Science, 62*(4), 771–804.

Friend, S. (2007). Fictional characters. *Philosophy Compass, 2*(2), 141–156.

Friend, S. (2020). The fictional character of scientific models. In P. Godfrey-Smith, & A. Levy (Eds.), *The scientific imagination. Philosophical and psychological perspectives* (pp. 102–127). New York: Oxford University Press.

Frigg, R. (2002). Models and representation: why structures are not enough. *Measurement in Physics and Economics Project Discussion Paper Series, DP MEAS 25/02*.

Frigg, R. (2003). *Re-presenting scientific represenation*. PhD Thesis: University of London.

Frigg, R. (2006). Scientific representation and the semantic view of theories. *Theoria, 55*(1), 49–65.

Frigg, R. (2010a). Models and fiction. *Synthese, 172*(2), 251–268.

Frigg, R. (2010b). Fiction and scientific representation. In R. Frigg, & M. Hunter (Eds.), *Beyond mimesis and convention: representation in art and science* (pp. 97–138). Berlin and New York: Springer.

Frigg, R. (2010c). Fiction in science. In J. Woods (Ed.), *Fictions and models: new essays* (pp. 247–287). Munich: Philiosophia Verlag.

Frigg, R. (2013). Clever fetishists *Art History, 36*(3), 665–669.

Frigg, R., Berkovitz, J., & Kronz, F. (2016). The ergodic hierarchy. In E. N. Zalta (Ed.), *The Stanford Encyclopedia of Philosophy*. URL = <https://plato.stanford.edu/archives/sum2016/entries/ergodic-hierarchy/>.

Frigg, R., Bradley, S., Du, H., & Smith, L. A. (2014). Laplace's demon and the adventures of his apprentices. *Philosophy of Science, 81*(1), 31–59.

Frigg, R., & Hartmann, S. (2020). Models in Science. In E. N. Zalta (Ed.), *The Stanford Encyclopedia of Philosophy*. URL = <https://plato.stanford.edu/archives/spr2020/entries/models-science/>.

Frigg, R., & Nguyen, J. (2016a). Scientific Representation. In E. N. Zalta (Ed.), *The Stanford Encyclopedia of Philosophy*. URL = <https://plato.stanford.edu/archives/win2018/entries/scientific-representation/>.

Frigg, R., & Nguyen, J. (2016b). The Fiction View of Models Reloaded. *The Monist, 99*, 225–242.

Frigg, R., & Nguyen, J. (2017a). Models and representation. In L. Magnani, & T. Bertolotti (Eds.), *Springer handbook of model-based science* (pp. 49–102). Dordrecht and Heidelberg: Springer.

Frigg, R., & Nguyen, J. (2017b). Scientific represenation is represenation as. In H.-K. Chao, & R. Julian (Eds.), *Philosophy of science in practice: Nancy Cartwright and the nature of scientific reasoning* (pp. 149–179). Cham: Springer.

Frigg, R., & Nguyen, J. (2017c). Of barrels and pipes: representation-as in art and science. In O. Bueno, G. Darby, S. French, & D. Rickles (Eds.), *Thinking about science, reflecting on art* (pp. 41–61). London: Routledge.

Frigg, R., & Nguyen, J. (2018). The turn of the valve: representing with material models. *European Journal for Philosophy of Science, 8*(2), 205–224.

Frigg, R., & Nguyen, J. (2019a). Of barrels and pipes: representation-as in art and science. In S. Wuppuluri (Ed.), *On art and science. Tango of an eternally inseparable duo* (pp. 181–202). Cham: Springer.

Frigg, R., & Nguyen, J. (2019b). Mirrors without warnings. *Synthese, Online First*. DOI:https://doi.org/10.1007/s11229-019-02222-9.

Frigg, R., & Votsis, I. (2011). Everything you always wanted to know about structural realism but were afraid to ask. *European Journal for Philosophy of Science, 1*(2), 227–276.

Frisch, M. (2014). Models and scientific representations or: who is afraid of inconsistency? *Synthese, 191*(13), 3027–3040.

Frisch, M. (2015). Users, structures, and representation. *The British Journal for the Philosophy of Science, 66*(2), 285–306.

Fumagalli, R. (2015). No learning from minimal models. *Philosophy of Science, 82*(5), 798–809.

Fumagalli, R. (2016). Why we cannot learn from minimal models. *Erkenntnis, 81*(3), 433–455.

Gallais, M. (2019). *Modèles scientifiques et objets théoriques. Essai d'épistémologie modale.* (Cahiers de Logique et d'Epistemologie). London: College Publications.

Gallegos, S. A. (2019). Models as signs: extending Kralemann and Lattman's proposal on modeling models within Peirce's theory of signs. *Synthese, 196,* 5115–5136.

Gaut, B. (2003). Art and knowledge. In J. Levinson (Ed.), *The Oxford handbook of aesthetics* (pp. 439–441). Oxford: Oxford University Press.

Gelfert, A. (2015). Symbol systems as collective representational resources: Mary Hesse, Nelson Goodman, and the problem of scientific representation. *Social Epistemology Review and Reply Collective, 4*(6), 52–61.

Gelfert, A. (2016). *How to do science with models: a philosophical primer* (SpringerBriefs in Philosophy). Switzerland: Springer

Gibson, J. (2008). Cognitivism and the arts. *Philosophy Compass, 3*(4), 573–589.

Giere, R. N. (1988). *Explaining science: a cognitive approach.* Chicago and London: University of Chicago Press.

Giere, R. N. (1994). No representation without representation. *Biology and Philosophy, 9*(1), 113–120.

Giere, R. N. (1996). Visual models and scientific judgement. In B. S. Baigrie (Ed.), *Picturing knowledge: historical and philosophical problems concerning the use of art in science* (pp. 269–302). Toronto: University of Toronto Press.

Giere, R. N. (1999). Using models to represent reality. In L. Magnani, N. J. Nersessian, & P. Thagard (Eds.), *Model-based reasoning in scientific discovery* (pp. 41–57). Dordrecht: Kluwer.

Giere, R. N. (2004). How models are used to represent reality. *Philosophy of Science, 71*(4), 742–752.

Giere, R. N. (2006). *Scientific perspectivism.* Chicago and London: University of Chicago Press.

Giere, R. N. (2009). Why scientific models should not be regarded as works of fiction. In M. Suárez (Ed.), *Fictions in science. Philosophical essays on modelling and idealization* (pp. 248–258). London: Routledge.

Giere, R. N. (2010). An agent-based conception of models and scientific representation. *Synthese, 172*(1), 269–281.

Glanzberg, M. (2018). Truth. In E. N. Zalta (Ed.), *The Stanford Encyclopedia of Philosophy.* URL = <https://plato.stanford.edu/archives/fall2018/entries/truth/>.

Glymour, C. (2013). Theoretical equivalence and the semantic view of theories. *Philosophy of Science, 80*(2), 286–297.

Godfrey-Smith, P. (2006). The strategy of model-based science. *Biology and Philosophy, 21*(5), 725–740.

Godfrey-Smith, P. (2009). Models and fictions in science. *Philosophical Studies, 143,* 101–116.

Godfrey-Smith, P. (2012). Metaphysics and the philosophical imagination. *Philosophical Studies, 160*(1), 97–113.

Godfrey-Smith, P. (2020). Models, fictions and conditions. In A. Levy, & P. Godfrey-Smith (Eds.), *The scientific imagination. Philosophical and psychological perspectives* (pp. 154–177). Cambridge: Cambridge University Press.

Gombrich, E. (1961). *Art and illusion.* Princeton: Princeton University Press.

Goodman, N. (1972). Seven strictures on similarity. In N. Goodman (Ed.), *Problems and projects* (pp. 437–446). Indianapolis and New York.

Goodman, N. (1976). *Languages of art* (2nd ed.). Indianapolis and Cambridge: Hackett.

Goodman, N. (1983). *Fact, fiction, and forecast* (4th ed.). Cambridge, MA: Harvard University Press.

Gräbner, C. (2018). How to relate models to reality? An epistemological framework for the validation and verification of computational models. *Journal of Artificial Societies and Social Simulation, 21*(3), nb. 8.

Grimm, S. R., Baumberger, C., & Ammon, S. (Eds.). (2017). *Explaining understanding: new perspectives from epistemology and philosophy of science.* New York and Abingdon: Routledge.

Grüne-Yanoff, T. (2009). Learning from minimal economic models. *Erkenntnis, 70*(1), 81–99.

Grüne-Yanoff, T. (2013). Appraising models nonrepresentationally. *Philosophy of Science, 80*(5), 850–861.
Grüne-Yanoff, T., & Schweinzer, P. (2008). The roles of stories in applying game theory. *Journal of Economic Methodology, 15*(2), 131–146.
Hacking, I. (1983). *Representing and intervening: introductory topics in the philosophy of natural science*. Cambridge: Cambridge University Press.
Halbach, V. (2014). *Axiomatic theories of truth* (2nd ed.). Cambridge: Cambridge University Press.
Hale, S. (1988). Spacetime and the abstract/concrete distinction. *Philosophical Studies, 53*(1), 85–102.
Halvorson, H. (2012). What scientific theories could not be. *Philosophy of Science, 79*(2), 183–206.
Halvorson, H. (2013). The semantic view, if plausible, is syntactic. *Philosophy of Science, 80*(3), 475–478.
Harris, T. (2003). Data models and the acquisition and manipulation of data. *Philosophy of Science, 70*(5), 1508–1517.
Hartmann, S. (1995). Models as a tool for theory construction: some strategies of preliminary physics. In W. E. Herfel, W. Krajewski, I. Niiniluoto, & R. Wojcicki (Eds.), *Theories and models in scientific processes (Poznan Studies in the Philosophy of Science and the Humanities 44)* (pp. 49–67). Amsterdam and Atlanta: Rodopi.
Hartmann, S. (1999). Models and stories in hadron physics. In M. Morgan, & M. Morrison (Eds.), *Models as mediators. Perspectives on natural and social science* (pp. 326–346). Cambridge: Cambridge University Press.
Hartmann, S. (2008). Modeling in philosophy of science. In M. Frauchiger, & W. K. Essler (Eds.), *Representation, evidence, and justification: themes from Suppes* (pp. 95–121), Lauener Library of Analytical Philosophy, Vol. 1. Frankfurt: Ontos.
Hartshorne, C., & Weiss, P. (Eds.). (1931–1935). *Collected papers of Charles Sanders Peirce. Volumes I - VI*. Cambridge, MA: Harvard University Press.
Hellman, G. (1989). *Mathematics without numbers: towards a modal-structural interpretation*. Oxford: Oxford University Press.
Hellman, G. (1996). Structuralism without structures. *Philosophia Mathematica, 4*(3), 100–123.
Hempel, C. G. (1970). On the "standard conception" of scientific theories. In M. Radner & S. Winokur (Eds.), *Minnesota studies in the philosophy of science* (Vol. 4, pp. 142–163). Minneapolis: University of Minnesota Press.
Hendry, R. F. (1998). Models and approximations in quantum chemistry. In N. Shanks (Ed.), *Idealization IX: idealization in contemporary physics* (pp. 123–142). Amsterdam: Rodopi.
Hesse, M. (1963). *Models and analogies in science*. London: Sheed and Ward.
Hodges, W. (1997). *A shorter model theory*. Cambridge: Cambridge University Press.
Howell, R. (1979). Fictional objects: how they are and how they aren't. *Poetics 8*, 129–177.
Hudetz, L. (2019). The semantic view of theories and higher-order languages. *Synthese, 196*, 1131–1149.
Hughes, R. I. G. (1997). Models and representation. *Philosophy of Science, 64*, S325–S336.
Hughes, R. I. G. (2010). *The Theoretical Practises of Physics: Philosophical Essays*. Oxford: Oxford Univeristy Press.
Illari, P. (2019). Mechanisms, models and laws in understanding supernovae. *Journal for General Philosophy of Science, 50*, 63–84.
Isaac, A. M. C. (2013). Modeling without representation. *Synthese, 190*(16), 3611–3623.
Isaac, A. M. C. (2019). The allegory of isomorphism. *AVANT. Trends in Interdisciplinary Studies, X*(3), 1–23.
Jebeile, J., & Kennedy, A. G. (2015). Explaining with models: the role of idealizations. *International Studies in the Philosophy of Science, 29*(4), 383–392.
Jhun, J., Palacios, P., & Weatherall, J. O. (2018). Market crashes as critical phenomena? Explanation, idealization, and universality in econophysics. *Synthese, 195*, 4477–4505.
Kalderon, M. E. (Ed.). (2005a). *Fictionalism in metaphysics*. Oxford: Oxford University Press.
Kalderon, M. E. (2005b). *Moral fictionalism*. Oxford: Clarendon Press.

Kennedy, A. G. (2012). A non representationalist view of model explanation. *Studies in History and Philosophy of Science, 43*(2), 326–332.

Ketland, J. (2004). Empirical adequacy and ramsification. *The British Journal for the Philosophy of Science, 55*(2), 287–300.

Khalifa, K. (2017). *Understanding, explanation, and scientific knowledge*. Cambridge: Cambridge University Press.

Khosrowi, D. (2020). Getting serious about shared features. *The British Journal for the Philosophy of Science, 71*(2), 523–546.

Kirkham, R. L. (1992). *Theories of truth: a critical introduction*. Cambridge, MA: MIT Press.

Klein, C. (2013). Multiple realizability and the semantic view of theories. *Philosophical Studies, 163*(3), 683–695.

Klein, U. (Ed.). (2001). *Tools and modes of representation in the laboratory sciences* (Boston Studies in the Philosophy of Science). Dordrecht and London: Kluwer Academic Publishers.

Knuuttila, T. (2005). Models, representation, and mediation. *Philosophy of Science, 72*(5), 1260–1271.

Knuuttila, T. (2011). Modelling and representing: An artefactual approach to model-based representation. *Studies in History and Philosophy of Science, 42*(2), 262–271.

Knuuttila, T. (2017). Imagination extended and embedded: artifactual versus fictional accounts of models. *Synthese, Online First. DOI:*https://doi.org/10.1007/s11229-017-1545-2.

Knuuttila, T., & Loettgers, A. (2017). Modelling as indirect representation? The Lotka–Volterra model revisited. *The British Journal for the Philosophy of Science, 68*(4), 1007–1036.

Kostić, D. (2019). Minimal structure explanations, scientific understanding and explanatory depth. *Perspectives on Science, 27*(1), 48–67.

Kralemann, B., & Lattmann, C. (2013). Models as icons: modeling models in the semiotic framework of Peirce's theory of signs. *Synthese, 190*(16), 3397–3420.

Kroes, P. (1989). Structural analogies between physical systems. *The British Journal for the Philosophy of Science, 40*(2), 145–154.

Kroon, F., & Voltolini, A. (2018). Fictional entities. In E. N. Zalta (Ed.), *The Stanford Encyclopedia of Philosophy* URL = <https://plato.stanford.edu/archives/win2018/entries/fictional-entities/>.

Kulvicki, J. (2006a). *On images: their structure and content*. Oxford: Oxford University Press.

Kulvicki, J. (2006b). Pictorial representation. *Philosophy Compass, 1*(6), 535–546.

Künne, W. (2003). *Conceptions of truth*. Oxford: Clarendon Press.

Kuypers, F. (1992). *Klassische Mechanik*. (3rd ed.). Weinheim: VHC.

Lamarque, P., & Olsen, S. H. (1994). *Truth, fiction, and literature*. Oxford: Clarendon Press.

Landry, E. (2007). Shared structure need not be shared set-structure. *Synthese, 158*(1), 1–17.

Landry, E. (Ed.). (2017). *Categories for the working philosopher*. Oxford: Oxford University Press.

Laplace, M. d. (1814). *A philosophical essay on probilities* (Dover Edition 1995). New York: Dover.

Laurence, S., & Margolis, E. (1999). Concepts and cognitive science. In S. Laurence, & E. Margolis (Eds.), *Concepts: core readings* (pp. 3–81). Cambridge, MA: MIT Press.

Lawler, I. (2019). Scientific understanding and felicitous legitimate falsehoods. *Synthese, Online First. DOI:*https://doi.org/10.1007/s11229-019-02495-0.

Lawler, I., & Sullivan, E. (2020). Model explanation versus model-induced explanation. *Foundations of Science, Online First. DOI:*https://doi.org/10.1007/s10699-020-09649-1.

Laymon, R. (1990). Computer simulations, idealizations and approximations. *Philosophy of Science. Proceedings of the Biennial Meeting of the Philosophy of Science Association. Part II: Symposia Papers*, 519–534.

Le Bihan, S. (2012). Defending the semantic view: what it takes. *European Journal for Philosophy of Science, 2*(3), 249–274.

Le Bihan, S. (2019). Partial truth versus felicitous falsehoods. *Synthese, Online First. DOI:*https://doi.org/10.1007/s11229-019-02413-4.

Leeson, R. (Ed.). (2000). *A. W. H. Phillips: collected works in contemporary perspective*. Cambridge: Cambridge University Press.

Leggett, D. (2013). Replication, re-placing and naval science in comparative context, c. 1868–1904. *The British Journal for the History of Science, 46*(1), 1–21.

Leinster, T. (2014). *Basic category theory* (Cambridge Studies in Advanced Mathematics). Cambridge: Cambridge University Press.

Leng, M. (2010). *Mathematics and reality*. Oxford: Oxford University Press.

Leonelli, S. (2016). *Data-centric biology: a philosophical study*. Chicago: University of Chicago Press.

Leonelli, S. (2019). What distinguishes data from models? *European Journal for Philosophy of Science, 9*(22), 1–27.

Leplin, J. (1980). The role of models in theory construction. In T. Nickles (Ed.), *Scientific discovery, logic, and rationality* (pp. 267–283). Dordrecht: Reidel.

Levy, A. (2012). Models, fictions, and realism: two packages. *Philosophy of Science, 79*(5), 738–748.

Levy, A. (2015). Modeling without models. *Philosophical Studies, 152*(3), 781–798.

Levy, A., & Currie, A. (2015). Model organisms are not (theoretical) models. *The British Journal for the Philosophy of Science, 66*(2), 327–348.

Lewis, D. (1978). Truth in fiction. In D. Lewis (Ed.), *Philosophical Papers, Volume I* (pp. 261–280). Oxford: Oxford University Press 1983.

Lichtenberg, A. J., & Liebermann, M. A. (1992). *Regular and chaotic dynamics* (2nd ed.). Berlin and New York: Springer.

Liu, C. (1999). Explaining the emergence of cooperative phenomena. *Philosophy of Science, 66 (Proceedings of the 1998 Biennial Meetings of the Philosophy of Science Association. Part I: Contributed Papers)*, S92–S106.

Liu, C. (2013). Deflationism on scientific representation. In V. Karakostas, & D. Dieks (Eds.), *EPSA11 Perspectives and Foundational Problems in Philosophy of Science* (pp. 93–102). Cham and Heidelberg: Springer.

Liu, C. (2014). Models, fiction, and fictional models. In G. Guo, & C. Liu (Eds.), *Scientific explanation and methodology of science* (pp. 107–127). Singapore: World Scientific.

Liu, C. (2015a). Re-inflating the conception of scientific representation. *International Studies in the Philosophy of Science, 29*(1), 41–59.

Liu, C. (2015b). Symbolic versus modelistic elements in scientific modeling. *Theoria, 30*(2), 287–300.

Liu, C. (2016). Against the new fictionalism: a hybrid view of scientific models. *International Studies in the Philosophy of Science, 30*(1), 39–54.

Lloyd, E. (1984). A semantic approach to the structure of population genetics. *Philosophy of Science, 51*(2), 242–264.

Lloyd, E. (1994). *The structure and confirmation of evolutionary theory*. Princeton: Princeton University Press.

Lopes, D. (2004). *Understanding pictures*. Oxford: Oxford University Press.

Luczak, J. (2017). Talk about toy models. *Studies in History and Philosophy of Modern Physics, 57*, 1–7.

Lutz, S. (2017). What was the syntax-semantics debate in the philosophy of science about? *Philosophy and Phenomenological Research, 95*(2), 319–352.

Lycan, W. G. (2000). *Philosophy of language: a contemporary introduction*. 2nd ed. (Routledge Contemporary Introductions to Philosophy). London: Routledge.

Lynch, M., & Woolgar, S. (1990). *Representation in scientific practice*. Cambridge, MA: MIT Press.

Mac Lane, S. (1998). *Categories for the working mathematician*. 2nd ed. (Graduate Texts in Mathematics). New York: Springer.

Machover, M. (1996). *Set theory, logic and their limitations*. Cambridge: Cambridge University Press.

MacKay, A. F. (1968). Mr. Donnellan and Humpty Dumpty on referring. *Philosophical Review, 77*(2), 197–202.

Magnani, L. (2012). Scientific models are not fictions: model-based science as epistemic warfare. In L. Magnani, & P. Li (Eds.), *Philosophy and cognitive science: western and eastern studies* (pp. 1–38). Berlin-Heidelberg: Springer-Verlag.

Mäki, U. (2009). MISSing the world. Models as isolations and credible surrogate systems. *Erkenntnis, 70*(1), 29–43.

Mäki, U. (2011). Models and the locus of their truth. *Synthese, 180*(1), 47–63.

Mandelbrot, B. B. (1982). *The fractal geometry of nature*. San Francisco: W.H.Freeman & Co Ltd.

Massimi, M. (2017). Perspectivism. In J. Saatsi (Ed.), *The Routledge handbook of scientific realism* (pp. 164–175). London and New York: Routledge.

Massimi, M. (2018). Perspectival modeling. *Philosophy of Science, 85*(3), 335–359.

Massimi, M., & McCoy, C. D. (Eds.). (2019). *Understanding perspectivism*. New York: Routledge.

McAllister, J. W. (1997). Phenomena and patterns in data sets. *Erkenntnis, 47*(2), 217–228.

McCloskey, D., N. (1990). Storytelling in economics. In C. Nash (Ed.), *Narrartive in culture. The uses of Storytelling in the sciences, philosophy, and literature* (pp. 5–22). London: Routledge.

McLoone, B. (2019). Thumper the infinitesimal rabbit: a fictionalist perspective on some "unimaginable" model systems in biology. *Philosophy of Science, 86*(4), 662–671.

Meinong, A. (1904). Über Gegenstandtheorie. In A. Meinong (Ed.), *Untersuchungen zur Gegenstandtheorie und Psychologie* (pp. 1–50). Leipzig: Barth.

Michaelson, E., & Reimer, M. (2019). Reference. In E. N. Zalta (Ed.), *The Stanford Encyclopedia of philosophy* (Spring 2019 Edition). https://plato.stanford.edu/archives/spr2019/entries/reference/.

Mitchell, S. D. (2002). Integrative pluralism. *Biology and Philosophy, 17*(1), 55–70.

Morgan, M. (2001). Models, stories and the economic world. *Journal of Economic Methodology, 8*(3), 361–384.

Morgan, M. (2004). Imagination and imaging in model building. *Philosophy of Science, 71*(4), 753–766.

Morgan, M. (2012). *The world in the model. How economists work and think*. Cambridge: Cambridge University Press.

Morgan, M., & Boumans, M. (2004). Secrets hidden by two-dimensionality: the economy as a hydraulic machine. In S. de Chadarevian, & N. Hopwood (Eds.), *Models: The Third Dimension of Science* (pp. 369–401). Stanford: Stanford University Press.

Morgan, M., & Morrison, M. (Eds.). (1999). *Models as mediators: perspectives on natural and social science*. Cambridge: Cambridge University Press.

Morreau, M. (2010). It simply does not add up: the trouble with overall similarity. *Journal of Philosophy, 107*(9), 469–490.

Morrison, M. (2008). Models as representational structures. In S. Hartmann, C. Hoefer, & L. Bovens (Eds.), *Nancy Cartwright's philosophy of science* (pp. 67–90), Routledge studies in the philosophy of science, Vol. 3. New York: Routledge.

Morrison, M. (2009). Fictions, representations, and reality. In M. Suárez (Ed.), *Fictions in science: philosophical essays on modeling and idealisation* (pp. 110–135). New York: Routledge.

Morrison, M. (2011). One phenomenon, many models: inconsistency and complementarity. *Studies in History and Philosophy of Science, 42*(2), 342–351.

Morrison, M. (2015). *Reconstructing reality: models, mathematics, and simulations*. New York: Oxford University Press.

Muller, F. A. (2011). Reflections on the revolution at stanford. *Synthese, 183*(1), 87–114.

Mundy, B. (1986). On the general theory of meaningful representation. *Synthese, 67*(3), 391–437.

Murphy, A. (2020). Towards a pluralist account of the imagination in science. Forthcoming in *Philosophy of Science*.

Nagel, E. (1961). *The structure of science*. London: Routledge and Keagan Paul.

Newlyn, W. T. (1950). The Phillips/Newlyn hydraulic model. *Yorkshire Bulletin of Economic and Social Research, 2*, 111–127.

Nguyen, J. (2016). On the pragmatic equivalence between representing data and phenomena. *Philosophy of Science, 83*(2), 171–191.

Nguyen, J. (2017). Scientific representation and theoretical equivalence. *Philosophy of Science, 84*(5), 982–995.

Nguyen, J. (2020). It's not a game: accurate representation with toy models. *The British Journal for the Philosophy of Science, 71*(3), 1013–1041.

Nguyen, J., & Frigg, R. (2017). Mathematics is not the only language in the book of nature. *Synthese, Online First. DOI:*https://doi.org/10.1007/s11229-017-1526-5.

Nguyen, J., Teh, N. J., & Wells, L. (2020). Why surplus structure is not superfluous. *The British Journal for the Philosophy of Science, 71*(2), 665–695.

Niiniluoto, I. (1988). Analogy and similarity in scientific reasoning. In D. H. Helman (Ed.), *Analogical reasoning: perspectives of artificial intelligence, cognitive science, and philosophy* (pp. 271–298). Dordrecht: Kluwer.

Niven, W. D. (1965). *The scientific papers of James Clerk Maxwell.* New York: Dover Publications.

Norton, J. (2003). Causation as folk science. *Philosophers' Imprint, 3*(4), 1–22.

Norton, J. (2008). The dome: an unexpectedly simple failure of determinism. *Philosophy of Science, 75*(5), 786–798.

Norton, J. (2012). Approximation and idealization: why the difference matters. *Philosophy of Science, 79*(2), 207–232.

O'Connor, C., & Weatherall, J. O. (2016). "Black holes, Black-Scholes, and prairie voles: an essay review of simulation and similarity, by Michael Weisberg. *Philosophy of Science, 83*(4), 613–626.

Odenbaugh, J. (2015). Semblance or similarity? Reflections on Simulation and Similarity. *Biology and Philosophy, 30*(2), 277–291.

Odenbaugh, J. (2018). Models, models, models: a deflationary view. *Synthese, Online First. DOI:*https://doi.org/10.1007/s11229-017-1665-8.

Osbeck, L. M., & Nersessian, N. J. (2006). The distribution of representation. *Journal for the Theory of Social Behaviour, 36*(2), 141–160.

Parker, M. W. (1998). Did Poincaré really discover chaos? *Studies in History and Philosophy of Modern Physics, 29*(4), 575–588.

Parker, W. (2015). Getting (even more) serious about similarity. *Biology and Philosophy, 30*(2), 267–276.

Parker, W. (2020). Model evaluation: An adequacy-for-purpose view. Forthcoming in *Philosophy of Science.*

Parsons, T. (1980). *Nonexistent objects.* New Haven: Yale University Press.

Perini, L. (2005a). The truth in pictures. *Philosophy of Science, 72*(1), 262–285.

Perini, L. (2005b). Visual representation and confirmation. *Philosophy of Science, 72*(5), 913–926.

Perini, L. (2010). Scientific representation and the semiotics of pictures. In P. D. Magnus, & J. Busch (Eds.), *New waves in the philosophy of science* (pp. 131–154). New York: Macmilan.

Pero, F., & Suárez, M. (2016). Varieties of misrepresentation and homomorphism. *European Journal for Philosophy of Science, 6*(1), 71–90.

Peschard, I. (2011). Making sense of modeling: beyond representation. *European Journal for Philosophy of Science, 1*(3), 335–352.

Phillips, A. W. (1950). Mechanical models in economic dynamics. *Economica, 17*(67), 283–305.

Pilyugin, S. Y. (1991). *Shadowing in dynamical systems.* Berlin, Heidelberg and New York: Springer.

Pincock, C. (2005). Overextending partial structures: idealization and abstraction. *Philosophy of Science, 72*(5), 1248–1259.

Pincock, C. (2012). *Mathematics and scientific representation.* Oxford: Oxford University Press.

Pincock, C. (2019). Concrete scale models, essential idealization and causal explanation. *The British Journal for the Philosophy of Science, Online First. DOI:*https://doi.org/10.1093/bjps/axz019.

Portides, D. (2005). Scientific models and the semantic view of theories. *Philosophy of Science, 72*(5), 1287–1289.

Portides, D. (2007). The relation between idealisation and approximation in scientific model construction. *Science & Education, 16*, 699–724.

Portides, D. (2010). Why the model-theoretic view of theories does not adequately depict the methodology of theory application. In M. Suárez, M. Dorato, & M. Rédei (Eds.), *EPSA Epistemology and Methodology of Science* (pp. 211–220). Dordrecht: Springer.

Portides, D. (2014). How scientific models differ from works of fiction. In L. Magnani (Ed.), *Model-based reasoning in science and technology: Theoretical and cognitive issues* (pp. 75–87). Berlin and Heidelberg: Springer.

Portides, D. (2017). Models and theories. In L. Magnani, & T. Bertolotti (Eds.), *Springer handbook of model-based science* (pp. 25–48). Dordrecht Heidelberg: Springer.

Potochnik, A. (2017). *Idealization and the Aims of Science*. Chicago and London: University of Chicago Press.

Poznic, M. (2016a). Make-believe and model-based representation in science: the epistemology of Frigg's and Toon's fictionalist views of modeling. *Theorema, 35*(3), 201–218.

Poznic, M. (2016b). Representation and similarity: Suárez on necessary and sufficient conditions of scientific representation. *Journal for General Philosophy of Science, 47*, 331–347.

Poznic, M. (2018). Thin versus thick accounts of scientific representation. *Synthese, 195*(8), 3433–3451.

Psillos, S. (1999). *Scientific realism: how science tracks truth*. London and New York: Routledge.

Purves, G. M. (2013). Finding truth in fictions: identifying non-fictions in imaginary cracks. *Synthese, 190*(2), 235–251.

Putnam, H. (1981). *Reason, truth, and history*. Cambridge: Cambridge University Press.

Putnam, H. (2002). *The collapse of the fact-value distinction*. Cambridge, MA: Harvard University Press.

Quine, W. V. O. (1969). *Ontological relativity and other essays*. New York: Columbia University Press.

Ramsey, J. L. (2006). Approximation. In S. Sarkar, & J. Pfeifer (Eds.), *The philosophy of science: an encyclopedia* (pp. 24–27). New York: Routledge.

Redhead, M. (2001). The intelligibility of the universe. In A. O'Hear (Ed.), *Philosophy at the New Millennium*. Cambridge: Cambridge University Press.

Regt, H. W. d. (2017). *Understanding Scientific Understanding*. New York: Oxford University Press.

Reiss, J. (2001). Natural economic quantities and their measurement. *Journal of Economic Methodology, 9*(2), 287–311.

Reiss, J. (2012a). The explanation paradox. *Journal of Economic Methodology, 19*(1), 43–62.

Reiss, J. (2012b). Idealization and the aims of economics: three cheers for instrumentalism. *Economics and Philosophy*, 363–383.

Reiss, J. (2013). Models, representation, and economic practice. In J.-H. Wolf, & U. Gähde (Eds.), *Models, simulations and the reduction of complexity* (pp. 107–116). Hamburg: DeGruyter.

Resnik, M. D. (1997). *Mathematics as a science of Ppatterns*. Oxford: Oxford University Press.

Reutlinger, A., Hangleiter, D., & Hartmann, S. (2018). Understanding (with) toy models. *The British Journal for the Philosophy of Science, 69*(4), 1069–1099.

Rice, C. (2018). Idealized models, holistic distortions, and universality. *Synthese, 195*(6), 2795–2819.

Rice, C. (2019). Models don't decompose that way: a holistic view of idealized models. *The British Journal for the Philosophy of Science, 70*(1), 179–208.

Rickart, C. E. (1995). *Structuralism and structure: a mathematical perspective*. Singapore: World Scientific Publishing.

Roberts, B. (2010). Group structural realism. *The British Journal for the Philosophy of Science, 62*(1), 47–69.

Rosen, G. (2020). Abstract objects. In E. N. Zalta (Ed.), *The Stanford Encyclopedia of Philosophy*. URL = <https://plato.stanford.edu/archives/spr2020/entries/abstract-objects/>.

Rosenblueth, A., & Wiener, N. (1945). The role of models in science. *Philosophy of Science, 12*(4), 316–321.

Rueger, A. (2005). Perspectival models and theory unification. *The British Journal for the Philosophy of Science, 56*(3), 579–594.

Rusanen, A.-M., & Lappi, O. (2012). An information semantic account of scientific models. In H. W. de Regt, S. Hartmann, & S. Okasha (Eds.), *EPSA Philosophy of Science: Amsterdam 2009* (pp. 315–328), The European Philosophy of Science Association Proceedings, Volume 1. Dordrecht and Heidelberg: Springer.

Russell, B. (1919/1993). *Introduction to mathematical philosophy*. London and New York: Routledge.

Saatsi, J. (2011a). Idealized models as inferentially veridical representations. In P. Humphreys, & C. Imbert (Eds.), *Models, simulations, and representations* (pp. 234–249). New York: Routledge.

Saatsi, J. (2011b). The enhanced indispensability argument: representational versus explanatory role of mathematics in science. *The British Journal for the Philosophy of Science, 62*(1), 143–154.

Saatsi, J. (2016). Models, idealisations, and realism. In E. Ippoliti, F. Sterpetti, & T. Nickles (Eds.), *Models and inferences in science* (pp. 173–189). Cham: Springer.

Salis, F. (2013). Fictional entities. In J. Branquinho, & R. Santos (Eds.), *Online Companion to Problems in Analytical Philosophy*. http://compendioemlinha.letras.ulisboa.pt.

Salis, F. (2014). Fictionalism. In J. Branquinho, & R. Santos (Eds.), *Online Companion to Problems in Analytical Philosophy*. http://compendioemlinha.letras.ulisboa.pt.

Salis, F. (2016). The nature of model-world comparisons. *The Monist, 99*(3), 243–259.

Salis, F. (2019). The new fiction view of models. *The British Journal for the Philosophy of Science, Online First. DOI:*https://doi.org/10.1093/bjps/axz015.

Salis, F., & Frigg, R. (2020). Capturing the scientific imagination. In P. Godfrey-Smith, & A. Levy (Eds.), *The scientific imagination. Philosophical and psychological perspectives* (pp. 17–50). New York: Oxford University Press.

Salis, F., Frigg, R., & Nguyen, J. (2020). Models and denotation. In C. Martínez-Vidal, & J. L. Falguera (Eds.), *Abstract objects: for and against* (pp. 197–219). Cham: Springer.

Schier, F. (1986). *Deeper in pictures: an essay on pictorial representation*. Cambridge: Cambridge University Press.

Shapiro, S. (1983). Mathematics and reality. *Philosophy of Science, 50*(4), 523–548.

Shapiro, S. (1997). *Philosophy of mathematics: structure and ontology*. Oxford: Oxford University Press.

Shapiro, S. (2000). *Thinking About Mathematics*. Oxford: Oxford University Press.

Shech, E. (2015). Scientific misrepresentation and guides to ontology: the need for representational code and contents. *Synthese, 192*, 3463–3485.

Shech, E. (2016). Fiction, depiction, and the complementarity thesis in art and science. *Monist, 99*, 311–332.

Shepard, R. N. (1980). Multidimensional scaling, tree-fitting, and clustering *Science, 210*(4468), 390–398.

Sismondo, S., & Chrisman, N. (2001). Deflationary metaphysics and the nature of maps. *Philosophy of Science (Proceedings), 68*, S38–49.

Sklar, L. (2000). *Theory and truth. Philosophical critique within foundational science*. Oxford: Oxford University Press.

Smith, L. A. (2007). *Chaos: a very short introduction*. Oxford: Oxford University Press.

Smith, P. (1998). *Explaining chaos*. Cambridge: Cambridge University Press.

Smolin, L. (2007). *The Trouble with Physics: The rise of String Theory, the Fall of a Science, and What Comes Next*. London: Allen Lane.

Spieler, O., Dingwell, D. B., & Alidibirov, M. (2004). Magma fragmentation speed: an experimental determination. *Journal of Volcanology and Geothermal Research, 129*(1–3), 109–123.

Spivak, M. (2006). *Calculus* (3rd ed.). Cambridge: Cambridge University Press.

Sprenger, J., & Hartmann, S. (2019). *Bayesian philosophy of science*. Oxford: Oxford University Press.

Stachowiak, H. (1973). *Allgemeine Modelltheorie*. Vienna and New York: Springer.

Sterelny, K., & Griffiths, P. E. (1999). *Sex and death: an introduction to philosophy of biology*. Chicago and London: University of Chicago Press.

Sterratt, D., Graham, B., Gilles, A., & Willshaw, D. (2011). *Principles of computational modelling in neuroscience*. Cambridge: Cambridge University Press.

Sterrett, S. G. (2002). Physical models and fundamental laws: using one piece of the world to tell about another. *Mind and Society, 5*(3), 51–66.

Sterrett, S. G. (2006). Models of machines and models of phenomena. *International Studies in the Philosophy of Science, 20*(1), 69–80.

Sterrett, S. G. (2020). Scale modeling. Forthcoming in D. Michelfelder, & N. Doorn (Eds.), *Routledge Handbook of Philosophy of Engineering*. London: Routledge.

Stevenson, M. (2011). The search for the fountain of prosperity. *Economia Politica, XXVIII*(1), 151–166.

Stich, S., & Warfield, T. (Eds.). (1994). *Mental representation. A reader*. Oxford: Blackwell.

Stoljar, D., & Damnjanovic, N. (2014). The deflationary theory of truth. In E. N. Zalta (Ed.), *The Stanford Encyclopedia of Philosophy*. URL = <https://plato.stanford.edu/archives/fall2014/entries/truth-deflationary/>.

Stolnitz, J. (1992). On the cognitive triviality of art. *The British Journal of Aesthetics, 32*(3), 191–200.

Strevens, M. (2008). *Depth*. Cambridge, MA: Harvard University Press.

Stuart, M. T. (2017). Imagination: a sine qua non of science. *Croatian Journal of Philosophy, XVII*(49), 9–32.

Stuart, M. T. (2020). The productive anarchy of scientific imagination. *Philosophy of Science, Online First. DOI:*https://doi.org/10.1086/710629.

Suárez, M. (2003). Scientific representation: against similarity and isomorphism. *International Studies in the Philosophy of Science, 17*(3), 225–244.

Suárez, M. (2004). An inferential conception of scientific representation. *Philosophy of Science, 71*(5), 767–779.

Suárez, M. (Ed.). (2009). *Fictions in science. Philosophical essays on modelling and idealization*. London and New York: Routledge.

Suárez, M. (2010). Scientific Representation. *Philosophy Compass, 5*(1), 91–101.

Suárez, M. (2015). Deflationary representation, inference, and practice. *Studies in History and Philosophy of Science, 49*, 36–47.

Suárez, M., & Cartwright, N. (2008). Theories: tools versus models. *Studies in History and Philosophy of Modern Physics, 39*, 62–81.

Suárez, M., & Pero, F. (2019). The representational semantic conception. *Philosophy of Science, 86*(2), 344–365.

Suárez, M., & Solé, A. (2006). On the analogy between cognitive representation and truth. *Theoria, 55*(1), 39–48.

Sugden, R. (2000). Credible worlds: the status of theoretical models in economics. *Journal of Economic Methodology, 7*(1), 1–31.

Sugden, R. (2009). Credible worlds, capacities and mechanisms. *Erkenntnis, 70*(1), 3–27.

Sullivan, E., & Khalifa, K. (2019). Idealizations and understanding: much ado about nothing? *Australasian Journal of Philosophy, 97*(4), 673–689.

Suppe, F. (1989). *The semantic conception of theories and scientific realism*. Urbana and Chicago: University of Illinois Press.

Suppes, P. (1969a). A comparison of the meaning and uses of models in mathematics and the empirical sciences. In P. Suppes (Ed.), *Studies in the methodology and foundations of science: selected papers from 1951 to 1969* (pp. 10–23). Dordrecht Reidel.

Suppes, P. (1969b). Models of data. In P. Suppes (Ed.), *Studies in the methodology and foundations of science: selected papers from 1951 to 1969* (pp. 24–35). Dordrecht: Springer.

Suppes, P. (1970). Set-theoretical structures in science. *Institute for Mathematical Studies in the Social Sciences*: Stanford University, Stanford.

Suppes, P. (2002). *Representation and invariance of scientific structures*. Stanford: CSLI Publications.

Swoyer, C. (1991). Structural representation and surrogative reasoning. *Synthese, 87*(3), 449–508.

Tal, E. (2017). Measurement in science. In E. N. Zalta (Ed.), *The Stanford Encyclopedia of Philosophy*. URL = <https://plato.stanford.edu/archives/fall2017/entries/measurement-science/>.

Taylor, J. H., & Vickers, P. (2017). Conceptual fragmentation and the rise of eliminativism. *European Journal for Philosophy of Science, 7*(1), 17–40.

Tegmark, M. (2008). The mathematical universe. *Foundations of Physics, 38*(2), 101–150.

Teller, P. (2001a). Twilight of the perfect model model. *Erkenntnis, 55*(3), 393–415.

Teller, P. (2001b). Whither constructive empiricism. *Philosophical Studies, 106*(1/2), 123–150.

Teller, P. (2009). Fictions, fictionalization, and truth in science. In M. Suárez (Ed.), *Fictions in science: philosophical essays in on modeling and idealization* (pp. 235–247). New York: Routledge.

Teller, P. (2018). Referential and Perspectival Realism. *Spontaneous Generations: A Journal for the History and Philosophy of Science, 9*(1), 151–164.

Thomasson, A. L. (1999). *Fiction and metaphysics*. New York: Cambridge University Press.

Thomasson, A. L. (2020). If models were fictions, then what would they be? In A. Levy, & P. Godfrey-Smith (Eds.), *The scientific imagination. Philosophical and psychological perspectives* (pp. 51–74). New York: Oxford University Press.

Thomson-Jones, K. (2005). Inseparable insight: reconciling formalism and cognitivism in aesthetics. *Journal of Aesthetics and Art Criticism, 63*(4), 375–384.

Thomson-Jones, M. (2010). Missing systems and face value practise. *Synthese, 172*(2), 283–299.

Thomson-Jones, M. (2011). Structuralism about scientific representation. In A. Bokulich, & P. Bokulich (Eds.), *Scientific structuralism* (pp. 119–141), Boston Studies in the Philosophy of Science, Vol. 281. Dordrecht: Springer.

Thomson-Jones, M. (2012). Modeling without mathematics. *Philosophy of Science, 79*(5), 761–772.

Thomson-Jones, M. (2020). Realism about missing systems. In A. Levy, & P. Godfrey-Smith (Eds.), *The scientific imagination. Philosophical and psychological perspectives* (pp. 75–101). New York: Oxford University Press.

Toon, A. (2010). Models as make-believe. In R. Frigg, & M. Hunter (Eds.), *Beyond mimesis and convention: representation in art and science* (pp. 71–96). Berlin Springer.

Toon, A. (2011). Playing with molecules. *Studies in History and Philosophy of Science, 42*, 580–589.

Toon, A. (2012a). *Models as make-believe. Imagination, fiction and scientific representation*. Basingstoke: Palgrave Macmillan.

Toon, A. (2012b). Similarity and scientific representation. *International Studies in the Philosophy of Science, 26*(3), 241–257.

Tversky, A. (1977). Features of similarity. *Psychological Review, 84*(4), 327–352.

Tversky, A., & Gati, I. (1978). Studies of similarity. In E. Rosch, & B. Lloyd (Eds.), *Cognition and categorization* (pp. 79–98). Hillside New Jersey Lawrence Elbaum Associates.

Ubbink, J. B. (1960). Model, description and knowledge. *Synthese, 12*(2–3), 302–319.

Vaihinger, H. (1911/1924). *The philosophy of 'as if': a system of the theoretical, practical, and religious fictions of mankind*. 1924 English translation, London: Kegan Paul.

van Fraassen, B. C. (1980). *The scientific image*. Oxford: Oxford University Press.

van Fraassen, B. C. (1981). Theory construction and experiment: an empiricist view. *Philosophy of Science. Proceedings of the 1980 Biennial Meetings of the Philosophy of Science Association. Part II: Symposia Papers, Vol. 2*, 663–677.

van Fraassen, B. C. (1985). Empricism in the philosophy of science. In P. M. Churchland, & C. A. Hooker (Eds.), *Images of science: essays on realism and empiricism with a reply from Bas C. van Fraassen* (pp. 245–308). Chicago and London: University of Chicago Press.

van Fraassen, B. C. (1989). *Laws and symmetry*. Oxford: Clarendon Press.

van Fraassen, B. C. (1991). *Quantum mechanics: an empiricist view.* Oxford: Oxford University Press.
van Fraassen, B. C. (1994). Interpretation of science; science as interpretation. In J. Hilgevoord (Ed.), *Physics and our view of the world* (pp. 169–187). Cambridge: Cambridge University Press.
van Fraassen, B. C. (1995). A philosophical approach to foundations of science. *Foundations of Science, 1*(1), 5–9.
van Fraassen, B. C. (1997). Structure and perspective: philosophical perplexity and paradox. In M. L. Dalla Chiara (Ed.), *Logic and scientific methods* (pp. 511–530). Dordrecht: Kluwer.
van Fraassen, B. C. (2002). *The empirical stance.* New Haven and London: Yale University Press.
van Fraassen, B. C. (2006). Representation: the problem for structuralism. *Philosophy of Science, 73*, 536–547.
van Fraassen, B. C. (2008). *Scientific representation: paradoxes of perspective.* Oxford: Clarendon Press.
van Fraassen, B. C. (2014). One or two gentle remarks about Hans Halvorson's critique of the semantic view. *Philosophy of Science, 81*(2), 276–283.
Verreault-Julien, P. (2019). Understanding does not depend on (causal) explanation. *European Journal for Philosophy of Science, 9*(18), 1–20.
Vickers, P. (2009). Can partial structures accomodate inconsistent science. *Principia, 13*(2), 233–250.
Vickers, P. (2016). Why Kirchhoff's approximation works. In K. Hentschel, & N. Y. Zhu (Eds.), *Gustav Robert Kirchhoff's treatise "on the theory of light rays"* (pp. 125–142). Singapore: World Scientific.
Vines, D. (2000). The Phillips Machine as a 'progressive' model. In R. Leeson (Ed.), *A. W. H. Phillips: collected works in contemporary perspective* (pp. 39–67). Cambridge: Cambridge University Press.
Vorms, M. (2011). Representing with imaginary models: formats matter. *Studies in History and Philosophy of Science, 42*(2), 287–295.
Vorms, M. (2012). Formats of representation in scientific theorising. In P. Humphreys, & C. Imbert (Eds.), *Models, simulations, and representations* (pp. 250–273). New York: Routledge
Wade, N. J., & Finger, S. (2001). The eye as an optical instrument: from camera obscura to Helmholtz's perspective. *Perception, 30*(10), 1157–1177.
Walton, K. L. (1990). *Mimesis as make-believe: on the foundations of the representational arts.* Cambridge, MA: Harvard University Press.
Warmbrōd, K. (1992). Primitive representation and misrepresentation. *Topoi, 11*(1), 89–101.
Wartofsky, M. W. (1979). *Models: representation and the scientific understanding.* Dordrecht: Reidel.
Weatherall, J. O. (2016a). Are Newtonian gravitation and geometrized Newtonian gravitation theoretically equivalent? *Erkenntnis, 81*(5), 1073–1091.
Weatherall, J. O. (2016b). Understanding gauge. *Philosophy of Science, 83*(5), 1039–1049.
Weatherall, J. O. (2018). Regarding the `hole arguement' *The British Journal for the Philosophy of Science, 2*(1), 329–350.
Weatherall, J. O. (2019a). Part 1: theoretical equivalence in physics. *Philosophy Compass, 14*(5), e12592.
Weatherall, J. O. (2019b). Part 2: theoretical equivalence in physics. *Philosophy Compass, 14*(5), e12591.
Webb, B. (2001). Can robots make good models of biological behaviour? *Behavioral and Brain Sciences, 24*(6), 1033–1050.
Weisberg, M. (2007a). Three kinds of idealization. *The Journal of Philosophy, 104*(12), 639–659.
Weisberg, M. (2007b). Who is a modeler? *The British Journal for the Philosophy of Science, 58*(2), 207–233.
Weisberg, M. (2012). Getting serious about similarity. *Philosophy of Science, 79*(5), 785–794.
Weisberg, M. (2013). *Simulation and similarity: using models to understand the world.* Oxford: Oxford University Press.

Weisberg, M. (2015). Biology and Philosophy symposium on Simulation and Similarity: using models to understand the world: response to critics. *Biology and Philosophy, 30*(2), 299–310.

Werndl, C. (2009). What are the new implications of chaos for unpredictability? *The British Journal for the Philosophy of Science, 60*, 195–220.

Werndl, C., & Frigg, R. (2015). Reconceptualising equilibrium in Boltzmannian statistical mechanics and characterising its existence. *Studies in History and Philosophy of Modern Physics, 49*, 19–31.

Wigner, E. (1960). The unreasonable effectiveness of mathematics in the natural sciences. *Communications on Pure and Applied Mathematics, 13*(1), 1–14.

Williamson, T. (2018). Model-building as philosophical method. *Phenomenology and Mind, 15*, 16–22.

Winsberg, E. (2009). A function for fictions: expanding the scope of science. In M. Suárez (Ed.), *Fictions in science: philosophical essays in on modeling and idealization* (pp. 179–191). New York: Routledge.

Wollheim, R. (1987). *Painting as an art* London: Thames and Hudson.

Woods, J. (2014). Against fictionalism. In L. Magnani (Ed.), *Model-based reasoning in science and technology: theoretical and cognitive issues* (pp. 9–42), Studies in applied philosophy, epistemology and rational ethics, Vol. 8. Berlin Heidelberg: Springer.

Woodward, J. (1989). Data and phenomena. *Synthese, 79*(3), 393–472.

Woodward, R. (2011). Truth in fiction. *Philosophy Compass, 6*(3), 158–167.

Woody, A. I. (2000). Putting quantum mechanics to work in chemistry: the power of diagrammatic pepresentation. *Philosophy of Science, 67*, S612–S627.

Woody, A. I. (2004). More telltale signs: what attention to representation reveals about scientific explanation. *Philosophy of Science, 71*(5), 780–793.

Worrall, J. (1989). Structural realism: the best of both worlds? *Dialectica, 43*(1–2), 99–124.

Yablo, S. (2014). *Aboutness*. Princeton: Princeton University Press.

Yablo, S. (2020). Models and reality. In P. Godfrey-Smith, & A. Levy (Eds.), *The scientific imagination. Philosophical and psychological perspectives* (pp. 128–153). New York: Oxford University Press.

Yaghmaie, A. (2012). Reflexive, symmetric and transitive scientific representations. PhilSci Archive, http://philsci-archive.pitt.edu/9454/.

Young, J. O. (2001). *Art and knowledge*. New York: Routledge.

Zalta, E. N. (1983). *Abstract objects: an introduction to axiomatic metaphysics*. Dordrecht: Reidel.

Name Index

A
Abell, C., 32
Achinstein, P., 108
Ackroyd, P., 35
Adams, E.W., 64
Ainsworth, P., 78
Aldana, E., 165
Alidibirov, M., 166
Ambrosio, C., 31
Ammon, S., xi
Angelico, F., 145
Ankeny, R.A., 16, 48, 166, 169
Anscombe, G.E.M., 41
Apostel, L., 39
Argyris, J.H., 182
Armstrong, D.M., 50, 79
Aronson, J.L., 36

B
Bailer-Jones, D.M., x, 3, 51, 85
Balaguer, M., 114
Balzer, W., 78
Barberousse, A., 121
Barcan Marcus, R., 180
Barnsley, M., 206
Barr, N., 161–165
Bartels, A., 61, 63
Batterman, R.W., xi, xii, 197
Baumberger, C., xi
Begg, D., 163
Bell, J., 66
Berkovitz, J., 210
Black, M., 36
Blumson, B., 32

Boesch, B., 28, 40–42, 46
Bogen, J., 71–73
Bohr, N., 123, 187
Bokulich, A., xi, xii, 113, 211
Bolinska, A., 3, 13, 101, 102
Boniolo, G., xiii
Boolos, G.S., 56
Borges, J.L., 12, 46
Boumans, M., 161, 171, 173
Brading, K., 60, 73
Bradley, S., 203
Brandom, R.B., 92
Budd, M., 54, 55
Bueno, O., 15, 28, 55, 58, 59, 61, 63–65, 78, 142, 191
Butterfield, J., 30, 80, 197
Byerly, H., 51, 61

C
Callender, C., 5, 8, 23–30, 90, 169
Carroll, L., 12, 29, 46
Cartwright, N., 38, 54, 91, 107, 108, 113, 204
Caulton, A., 69
Chakravartty, A., x, xi, 51, 93, 94
Chang, H., xi, xii
Chrisman, N., 175
Churchill, W., 36, 37, 149, 206
Cohen, J., 5, 8, 23–30, 90, 169
Colyvan, M., 7, 15, 142, 191
Contessa, G., 2–4, 83, 84, 86, 91, 93, 95–98, 100–102, 114, 125, 137, 171
Cornell Way, E., 36
Crittenden, C., 116
Cummins, R., 49

Curran, D., 14
Currie, A., 166, 210
Currie, G., 109, 119, 134, 210, 212
Cyr, S., 166

D
Da Costa, N.C.A., 51, 57, 58, 61, 65
Damnjanovic, N., 89
Danto, A., 143, 169
Dardashti, R., xi, 166
Davies, D., 108
de Chadarevian, S., 15, 166
Decock, L., 44
de Donato Rodriguez, X., 92
Degas, E., 205
Demopoulos, W., 78
Denis, M., 2, 3, 205
Dewar, N., 69
Dickens, C., 211
Díez, J., xii, 83, 99–101
Dingwell, D.B., 166
Donnellan, K.S., 29
Dornbusch, R., 163
Dostoevsky, F., 107, 211
Douven, I., 44
Downes, S.M., xii, 43, 51, 54
Doyle, Y., xi
Du, H., 203
Ducheyne, S., 24, 37, 83, 102, 103
Dummett, M., 58

E
Eco, U., 212
Eddon, M., 167
Egan, S., xi
Einstein, A., 107
Elgin, C.Z., xi, 8, 86, 134, 137–140, 143–148, 150–153, 155, 156, 159, 174, 181, 213
Elkins, J., xii, 206
Enderton, H.B., 57, 66
Evans, G., 109, 119

F
Faust, G., 182
Fine, A., 113
Finger, S., 166
Fischer, S., 163
Fischli, 207, 212
Fletcher, S.C., 68
Frege, G., 179

French, S., 8, 28, 51, 54, 55, 57–61, 63–65, 78, 80
Friend, S., 116–121, 125
Frigg, R., x, 2–4, 8, 9, 12, 16, 27, 35, 62, 73, 75, 76, 78, 103, 107, 109, 110, 116, 119, 121, 123, 128, 169, 175, 176, 203, 210
Frisch, M., 27, 77
Fumagalli, R., xii

G
Gallais, M., 116
Gallegos, S.A., 31
García Márquez, G., 212
Gati, I., 36
Gaut, B., 209
Gelfert, A., 28, 152
Gibson, J., 209
Giere, R.N., xi, 14, 27, 33, 37–42, 48–51, 107, 132, 133, 135, 169, 184, 204
Gilles, A., 166
Gillray, J., 145
Glanzberg, M., 88
Glymour, C., 57, 60
Godfrey-Smith, P., 7, 33, 54, 114, 116, 123, 125, 130
Gombrich, E., 154
Goodman, N., 11, 35, 36, 38, 44, 45, 79, 86, 137–139, 143–147, 149, 150, 152, 213
Gräbner, C., 190
Graham, B., 166
Graham, N., xi
Griffiths, P.E., 12
Grimm, S.R., xi
Grüne-Yanoff, T., xii, 108

H
Haase, M., 182
Hacking, I., 16, 48
Halbach, V., 89
Hale, S., 48
Halvorson, H., 57, 60, 68
Hangleiter, D., xi
Harré, R., 36
Harris, T., 70
Hartmann, S., x, xii, 7, 107, 210
Hartshorne, C., xii
Hellman, G., 58, 79
Hendry, R.F., 103
Hesse, M., 45, 46, 59, 176
Hodges, W., 56, 57, 66

Horwich, P., 89, 90
Howell, R., 117
Hudetz, L., 56
Hughes, R.I.G., 49, 83, 137–144, 150, 170

I
Ih, K.-D., 166
Illari, P., xi
Isaac, A., xii, 18, 55

J
Jebeile, J., x
Jeffrey, R.C., 56
Jhun, J., xii

K
Kafka, F., 211
Kalderon, M.E., 114
Kennedy, A.G., xi, xii, 211
Ketland, J., 78
Keynes, J.M., 163, 167, 169
Khalifa, K., xi
Khosrowi, D., 47
Kirkham, R.L., 88
Klein, C., 51
Klein, U., 16
Knuuttila, T., xii, 15, 33, 125, 132, 133, 211
Kostić, D., xi
Kralemann, B., 31
Kripke, S., 180
Kroes, P., 62
Kronz, F., 210
Kroon, F., 116
Kulvicki, J., 146, 154, 206
Künne, W., 88
Kuypers, F., 112

L
Ladyman, J., 57, 61, 65, 80
Lamarque, P., 109, 123
Landry, E., 60, 68, 73
Laplace, M.d., 201
Lappi, O., 39
Lattmann, C., 31
Laurence, S., 11
Lawler, I., xi, 151
Laymon, R., 103, 176
Le Bihan, S., xi, 51
Lee, H., 134

Leeson, R., 163
Leggett, D., 16, 166
Leibniz, G.W., 209
Leinster, T., 68
Leng, M., 114, 194
Leonelli, S., 16, 48, 70, 166, 169
Leplin, J., xii
Levy, A., 16, 116, 126–128, 130, 146, 166
Lewis, D., 12, 29, 117
Lichtenberg, A.J., 203
Liebermann, M.A., 203
Liu, C., 4, 33, 103, 113, 132
Lloyd, E., 54, 61, 66
Loettgers, A., 33, 125, 212
Lopes, D., 32, 43, 154
Luczak, J., xii, 13, 211
Ludwig, P., 121
Lutz, S., 57, 60
Lycan, W.G., 179
Lynch, M., 14

M
Machover, M., 56, 57, 66
MacKay, A.F., 29
Mac Lane, S., 68
Magnani, L., 132, 133
Magritte, R., 101
Mäki, U., 41, 42
Mandelbrot, B.B., 103, 181, 195, 204–207
Margolis, E., 11
Martin, G.R.R., 145
Massimi, M., xi, 184
Maxwell, J.C., 13, 21, 38, 107, 131
McAllister, J.W., 73
McCloskey, D.N., 107
McCoy, C.D., xi
McLoone, B., 110
Meinong, A., 117
Michaelson, E., 30
Mill, J.S., 180
Mitchell, S.D., xi
Moore, G.E., 73
Morgan, M., xii, 4, 54, 107, 159, 161, 171, 173, 204
Morreau, M., 43
Morrison, M., xi, xii, 2, 4, 12, 28, 54, 113, 131, 204
Moulines, C.U., 78
Muller, F.A., 63, 71
Mundy, B., 61, 66
Murphy, A., 110

N
Nersessian, N.J., 42
Newlyn, W.T., 9, 161, 208
Newton, I., ix, 2, 13, 18, 21, 32, 34, 37, 49, 79, 105, 118, 122, 124, 133, 187, 195, 199, 202, 203, 208, 212
Nguyen, J., vi, xi, 16, 58, 62, 65, 68, 69, 73, 101, 103, 128
Niiniluoto, I., 46
Niven, W.D., 107
Norton, J.D., x, 13, 103, 210

O
O'Connor, C., 34
Odenbaugh, J., 49, 134
Olsen, S.H., 109, 123
Orwell, G., 134, 208
Osbeck, L.M., 42

P
Palacios, P., xii
Parker, M.W., 203
Parker, W., 12, 46
Parsons, T., 117
Perini, L., xii
Pero, F., 51, 66, 67, 84
Peschard, I., xii, 211
Phidias, 145
Phillips, A.W., 161, 168, 174
Pilyugin, S.Y., 203
Pincock, C., 67, 76, 133, 175
Plato, 15
Poincaré, H., 202, 203
Portides, D., x, xii, 51, 77, 131, 132, 204
Potochnik, A., xi
Pourbus, F., 207
Poznic, M., 4, 13, 36, 128
Psillos, S., x
Purves, G.M., 113
Putnam, H., 36, 37, 39, 61, 78, 206

Q
Quine, W.V.O., 44

R
Ramsey, F., 88
Ramsey, J.L., 103
Redhead, M., 57, 61, 65
Regt, H.W.d., xi
Reimer, M., 29, 179
Reiss, J., x–xii, 42
Remarque, E.M., 211
Resnik, M.D., 52, 58, 79
Reutlinger, A., xi
Rice, C., xii, 129
Rickart, C.E., 56
Roberts, B.W., 60
Roberts, J., 145
Robertson, D., 163, 169
Rosen, G., 48
Rosenblueth, A., 16
Rowling, J.K., 115
Rueger, A., xi
Rusanen, A.-M., 39
Russell, B., 56, 179

S
Saatsi, J., x, xi, 15, 59
Salis, F., 107, 110, 113, 116, 123–125, 189
Schier, F., 154
Schweinzer, P., 108
Shapiro, S., 15, 52, 58, 79
Shech, E., 94, 98, 100, 101
Shepard, R.N., 44
Sismondo, S., 175
Sklar, L., 107
Smith, L.A., 106, 203
Smith, P., 203
Smolin, L., 121
Solé, A., 85, 90, 92, 93
Spieler, O., 166
Spivak, M., 19
Sprenger, J., 7
Stachowiak, H., 2
Sterelny, K., 12
Sterratt, D., 166
Sterrett, S.G., 166, 175
Stevenson, M., 165
Stich, S., 12
Stoljar, D., 89
Stolnitz, J., 209
Strevens, M., xi
Stuart, M.T., 110
Suárez, M., 2, 4, 8, 12, 27, 35, 37, 39, 40, 42, 51, 66, 67, 76, 83–90, 92, 93, 95, 113, 137, 142, 169, 204
Sugden, R., 108
Sullivan, E., xi
Suppe, F., 51

Suppes, P., 51, 57, 70, 74
Swoyer, C., 3, 27, 61, 66, 169

T
Tal, E., xii
Taylor, J.H., xi
Tegmark, M., 80, 81
Teh, N.J., xi, 65, 69
Teller, P., 44, 46, 50, 51, 73, 133, 171
Thébault, K.P.Y., xi, 168
Thomasson, A.L., 125, 127
Thomson-Jones, K., 212
Thomson-Jones, M., 15, 49–51, 58, 108, 109, 118, 127, 128
Tolstoy, L., 112, 133, 212
Toon, A., 27, 28, 39, 49, 126, 128–131, 168
Turner, W., 34
Tversky, A., 36, 44

U
Ubbink, J.B., 61
Uccello, P., 145

V
Vaihinger, H., 107
van Fraassen, B.C., 8, 27, 39, 51, 53, 57, 61, 62, 64, 65, 70, 71, 73, 74, 77, 137, 155, 204
Vargas Llosa, M., 211
Velázquez, D., 37, 84
Vernasca, G., 163
Verreault-Julien, P., xi
Vickers, P., xi, 59, 63
Vines, D., 159, 161, 163, 165
Voltaire, 209
Voltolini, A., 116
Vonnegut, K., 211
Vorms, M., 17, 212
Votsis, I., 78

W
Wade, N.J., 166
Walton, K.L., 108, 109, 118–121, 123, 124, 126
Warfield, T., 12
Warmbrōd, K., xii
Wartofsky, M.W., 26, 39
Weatherall, J.O., 34, 65, 68, 69
Webb, B., 166
Weisberg, M., xii, 4, 13, 32–34, 36, 44–48, 126, 133, 134, 146, 148, 166, 171, 190, 195, 210
Weiss, P., xii, 207, 212
Weizsäcker, C.F., 9
Werndl, C., 203, 210
Wiener, N., 15
Wigner, E., 15
Williamson, T., 7
Winsberg, E., 131
Wittgenstein, L., 73
Wollheim, R., 154
Woods, J., 113
Woodward, J., 71–73
Woodward, R., 116
Woody, A.I., xi, 9
Woolgar, S., 14
Worrall, J., 59
Wright, C., 89, 92

Y
Yablo, S., 128
Yaghmaie, A., 35, 36
Young, J.O., 154

Z
Zalta, E.N., 117
Zamora Bonilla, J., 92
Zola, É., 211

Subject Index

A
Abstract, 10, 14, 16, 33, 48–50, 57, 76, 77, 89–95, 99, 115, 117, 125, 134, 142, 146, 153, 179
Abstraction, 97, 147, 159, 178, 213
Accuracy, 11–13, 21, 26, 31, 40, 42–47, 52, 64–69, 72, 98, 128, 141, 151, 180, 190
Action, intentional, 41
Analogy, x, 28, 45, 74, 87–89, 92, 100, 105, 108, 115, 116, 124, 128, 129, 175, 176
Antirealism, 113, 117, 121, 124
Antirealist, 14, 113, 121, 124, 183, 186, 189
Aristotelian, 97
Art, xiii, 81, 84, 109, 134, 138, 143, 179, 185, 205–207, 209, 210, 212, 213
Artefacts, xii, 7, 62, 125
Audience, 27, 30, 41, 42, 135, 155

B
Biconditional, 3, 4, 6, 10, 20, 66, 85, 90
Blank, 3, 4, 6, 9, 20, 34, 55, 91, 175, 176, 179

C
Caenorhabditis elegans, 154, 169
Caricature, 74, 137, 145, 149, 152, 155
Carrier, 2, 26, 37, 52, 85, 105, 137, 167, 185
Category theory, 65, 68, 69
Characterisation, 16, 46, 48, 59, 72, 85, 89, 93–95, 169
 functional, 93–95
 inherent, 93, 94
Codomain, 52, 60, 65
Coherentist, 88
Concepts, classical theory of, 210
Construal, 105, 134, 171, 190, 195
Constructivist, 14
Correspondence theory, 88

D
Data, 70–74, 193
DDI, 83, 137, 139–142
Deflationary, 83, 85–90, 92, 95, 96, 142
Deflationism, 85, 87–93, 141
Deflationist, 87
DEKI, v, xiii, 159–213
Demarcation, 2–8, 10, 18, 20, 23–25, 31–34, 52–55, 83–84, 96, 137–139, 180, 185, 204–205, 210
Denotation, 28–29, 38, 41, 85, 86, 124, 125, 139–146, 149, 151, 159, 171, 176–180, 189
Description of a missing system, 106, 107, 117, 126, 127
 See also Missing system
Directionality, 3, 21, 26, 35, 39, 64, 91, 96, 124–125, 142, 151, 180, 189
Disquotation, 89
Distortions, 63, 67, 128
Domains, ix, 52–53, 56–58, 60, 65, 66, 74, 75, 79, 80, 99, 113, 138, 139, 167, 171, 177
Dynamics, 28, 32, 140–142, 147, 159, 169, 179, 187, 203, 211

E
Embedding, 52, 61, 65, 69
Equivalence, 58, 62, 68, 107

© Springer Nature Switzerland AG 2020
R. Frigg, J. Nguyen, *Modelling Nature: An Opinionated Introduction to Scientific Representation*, Synthese Library 427,
https://doi.org/10.1007/978-3-030-45153-0

Exemplification, 79, 143, 147–149, 152–155, 159, 172–173, 176
Explanation, xi, xii, 14, 27, 39, 41, 54, 60, 88, 89, 94, 129, 215
Extensionalisation, 192, 194, 195

F
Faithful, 101, 178, 207
Falsehoods, 103
Falsity, 89, 109, 110, 115, 132, 133
Fiat, xiii, 23–30
Fiction, xiii, 50, 105–135, 186, 188, 189, 194, 199–201, 209–212
Frictionless plane, 128, 131, 154, 199
Function, denotative, 142

G
Game of make-believe, 119–127, 187–188, 194
Genre, 108, 109, 146, 154, 155, 166, 212
God, 145, 156, 209
Griceanism, xiii, 23–30, 40
Group theory, 119

H
Handling, 16, 17, 21, 49, 50, 58, 87, 115, 116, 118, 121, 122, 125, 127
Homomorphism, 51, 61, 66, 67

I
Idealisations, x, 21, 67, 102, 103, 108, 128, 129, 139, 157, 192, 199
Imagination, 50, 105, 107–112, 114, 120, 124, 127, 130, 132, 188, 189
Imagined, 54, 105, 107, 110, 114, 119, 120, 122, 127, 187–190
Imputation, 149, 151, 174–176, 178, 192, 203
Inaccuracy, 87
Individual, 60, 69, 75, 79, 99, 100, 132, 138, 206
Inferences, 4, 17, 25, 26, 28, 42, 83–87, 90, 91, 93–99, 101, 118, 120, 127, 183, 212
Inferentialism, xiii, 83, 85–87, 90–103, 142
Inferentialist, 84, 85, 90–93, 95
Infidelity, 105, 109, 111–114
Interpretation, 11, 52, 93, 134, 139–142, 159, 185
Isomorphic, 60–65, 69, 71, 72, 75, 77, 78, 80, 81
Isomorphism, 25, 26, 29, 30, 35, 51, 54, 55, 61–66, 69, 70, 73, 78–80, 86, 92, 93, 96

J
Jesus, 156

K
Key, 73, 90, 93, 149, 159, 174–186, 189–192, 195–205, 207–209, 213

L
Limit, 20, 65, 105, 113, 176, 195–205, 207, 214
Literature, v, xi, xii, xiii, 8, 15, 41, 43, 44, 49, 56, 58, 68, 70, 79, 93, 94, 97, 108, 116, 117, 119, 125, 184, 208–212
Litmus, xii, 103, 157, 181, 185, 204, 207

M
Make-believe, 121–125, 188
Maps, 4, 12, 18, 46, 65, 67, 68, 81, 84, 138, 145, 156, 157, 168, 170, 175, 176, 179, 183, 191, 195, 205, 210
Mathematics, applicability of, 14, 15, 22, 27, 50, 52, 77, 80, 96, 97, 127, 141, 153, 180, 185, 190–195
Methane, 75–77
Misrepresentation, 1, 12–14, 21, 26, 37–39, 42, 63, 64, 87, 97–103, 127, 141, 150, 151, 174, 180, 192
Missing systems, 106, 107, 111, 117, 126, 127 *See also* Description of a missing system
Model-land, 106
Model organism, 48, 166, 169, 180
Model-system, 17, 106, 107, 114, 118, 123, 124, 126, 190
Morphism, 63, 67–70, 81, 133

N
Naturalistic, 39, 40
Newman's theorem, 75, 78, 79
Newtonian, 3, 15, 16, 25, 28, 35, 48, 53, 62, 67, 122, 124, 128, 150, 186, 202, 212
Nihilism, 109, 110

O
Ontology, 16, 21, 50, 51, 58, 59, 79, 81, 87, 115, 116, 118, 121, 123–125, 133, 186

P
Pattern, 3, 45, 72, 79, 147, 153, 159, 202, 206
Perspectivalism, 184

Subject Index 241

Perspectivism, xi
Phillips–Newlyn machine, 15, 48, 54, 159, 160, 162–173, 177–179, 183, 185, 187, 195, 205
Platonist, 79
Pluralism, xi, 9
Pragmatic, 8, 25–27, 29, 42, 64, 73, 74, 102, 113, 170
Pretence, 105, 108, 118–122, 124–127, 186, 188, 194
Principle of generation, 119, 120, 122, 123, 126, 134, 187, 188, 194, 199
Prop, 119–123, 126, 127, 187, 188
Property attribution, 17, 116, 123

R

Realism, x, 14, 59, 80, 113, 117, 125, 159, 183, 211, 213
Realist, x, 14, 80, 113, 117, 121, 153, 186, 189, 209
Reinflate, 83
Relation, 2, 23, 32, 53, 83, 115, 141, 189
  partial, 58, 59
  total, 58
Representation, 1, 23, 31, 53, 83, 105, 139, 160, 186
  direct, 33–35, 53, 55, 126–130, 204
  epistemic, 3, 23, 31, 54, 84, 105, 137, 161, 185
  indirect, 32–35, 48, 53, 55, 58, 61, 126
  mental, 12, 24, 49
  visual, xii, 9, 32, 204
Representational force, 85, 86, 90–92, 142
Representation-as, 83, 137–157, 159, 174–176, 178, 182, 204, 213
Resemblance, 31, 38, 41, 63, 139, 155

S

San Francisco Bay, 32, 47, 48, 54, 166, 167, 169, 171
Scale, 9, 106, 145, 156, 171, 174, 175, 195
Semantic, 17, 39, 43, 51, 52, 57–59, 61, 68, 77, 79, 98, 138, 150, 170, 204
Sequence, 41, 196–201, 203, 208, 213
Signature, 57, 58, 66
Similarity, xiii, 25, 26, 29–50, 53–55, 61–64, 83, 86, 92, 93, 126, 128, 131, 134, 137, 139, 141, 148, 155, 175, 181, 182, 185, 205, 206
  weighted feature matching, 44, 45
Stereotypes, 11
Stipulation, 24–30, 85, 139, 149

Stipulative, 23–30
Structuralism, 53, 54, 60–64, 77, 79, 190
Structuralist, xiii, 51–81, 96, 97, 141, 186, 190, 191
Structure, 1, 28, 43, 53, 87, 130, 137, 161, 186
  generating description, 76, 77, 79, 192–194
  mathematical, 16, 50, 53, 56, 58, 62, 68, 69, 73, 79–81, 87, 133, 189–191, 193–195
  partial, 54, 58–61, 63–65, 67
  set-theoretic, 56, 67, 70, 186
  total, 59
Style, 1, 9–12, 14, 17, 20, 26, 32, 35, 40, 42–47, 52, 64–69, 87, 97, 127, 141, 151, 170, 180, 208, 210, 212
Surrogative reasoning, 1–8, 18, 21, 26–29, 31, 42, 43, 52, 83, 86, 91, 92, 96, 101, 127–129, 141, 151, 174, 180
Symmetry, 14, 36, 38
Syntax, 57

T

Targetless, 12–14, 21, 26, 130, 142, 144, 146, 151, 153, 180, 210
Targets, 1, 24, 31, 53, 83, 111, 137, 159, 185
Theory, 2, 24, 33, 53, 86, 107, 139, 163, 188
Thought experiment, 36, 62, 108, 139
Toy models, xii
Truth, 17, 52, 89, 110, 149, 167, 195
  approximate, x
  partial, 128
Two-body system, 124, 187, 195

U

Underdetermination, 77, 78, 214
Understanding, v, x, xi, 1, 12, 16, 21, 30, 43, 49, 50, 77, 93–95, 106, 108, 109, 115, 128, 132, 133, 151, 156, 166, 173, 175, 183, 187, 194, 195, 203, 205, 206, 210, 213
Universal, ante rem, 79

V

Vehicles, 16, 24–26, 28, 33, 34, 41, 86, 96, 155

Z

Z-representation, 143–157, 159, 166–171, 174, 178–180, 182, 185, 186, 189, 190, 205, 206

CPSIA information can be obtained
at www.ICGtesting.com
Printed in the USA
LVHW082142131020
668754LV00014B/155